The Complete Book of HOME FREEZING

By Hazel Meyer

HAZEL MEYER'S FREEZER COOK BOOK

THE COMPLETE BOOK OF HOME FREEZING

THE COMPLETE BOOK OF

drawings by ARTHUR C. COLLINS

By HAZEL MEYER

HOME FREEZING

REVISED EDITION

J. B. LIPPINCOTT COMPANY
Philadelphia and New York
1970

FOURTH PRINTING, REVISED EDITION

ISBN-0-397-00639-X

LIBRARY OF CONGRESS CATALOG CARD NUMBER 71-139808

PRINTED IN THE UNITED STATES OF AMERICA

To the memory of my father,
who loved his family's enjoyment
of the picture-book meals
he so often prepared

Contents

CHARTS

ILLUSTRATIONS

TABLES

Acknowledgments

I want to thank the many helpful people who patiently suffered my impertinent questions, especially the officials, technicians and engineers of American food companies, home freezer manufacturers and packaging material suppliers, who took the time and trouble to furnish information which they themselves had won only through diligent research and costly development.

My particular gratitude goes to Mr. George C. Cook, head of the frozen food educational departments at Long Island Agricultural and Technical Institute, for reading this book in manuscript form with an expert's critical eye on its accuracy and usefulness.

Special acknowledgment is also due to the Bureau of Human Nutrition and Home Economics of the United States Department of Agriculture; the New York State Agricultural Experiment Station at Cornell University; the Agricultural Extension Service of Iowa State College; the Agricultural Experiment Station and Cooperative Extension Service of Michigan State College; the University of Maryland Extension Service; the National Live Stock and Meat Board; the United States Department of the Interior, Fish and Wildlife Service; the game conservation commissions of the forty-eight states; the Food and Agriculture Organization of the United Nations, and the editorial staffs of the trade journals and magazines serving the food industries of the nation.

Foreword

When the first edition of this book was in its planning stage almost twenty years ago, the purchase of a home freezer represented for most families a decision to be undertaken only after long and careful thought.

A freezer was once an entirely new kind of major appliance to many average households, and its initial cost, running into a few hundred dollars, was an important budget consideration in family discussions about a desirable but untried novelty.

During those twenty years, the freezer logically and inevitably moved into the familiar daily lives of American homes, whether as a separate appliance or in combination with a new-model refrigerator. While it still represents a fair-sized price tag, the economies of living with a freezer are perhaps even more pertinent toward the last quarter of this century than they were in its middle years. Food does not seem to be getting any cheaper.

The economies that come with proper use of a home freezer may be realized a few cents or a few dollars at a time, but—now as then—it does not take long before they add up to the amount invested for the appliance itself.

For example, a freezer owner can buy ice cream by the gallon, saving from twenty-five to forty-five cents on each quart formerly bought at the corner drug store. If a family consumes ten gallons of ice cream a year, which is less than one serving per week to each of a family of four, a ten to eighteen dollar saving can be chalked up for ice cream alone.

An economy-minded freezer family decides to buy a side of beef, a whole hog, a quarter of veal and a young lamb instead of getting table cuts as needed from a local butcher. Depending on the size of both family and freezer, and the amount of meat purchased, savings can mount up to 30 per cent of former costs.

A small flock of chickens, a modest vegetable patch and some berry bushes and perhaps a fruit tree or two in the back yard can add approximately two hundred dollars to savings on poultry, eggs, vegetables and desserts.

Even without a chicken coop or a garden, buying a dozen chickens or a few turkeys when prices are down can save up to a dollar a bird. If the family likes chicken enough to eat one every other week, the total saving on the main course of Sunday dinner alone is twenty-five dollars.

Bushels of beans and boxes of peaches bought in season can mean many more dollar savings, and so can buying case lots of already frozen and packaged foods from a nearby supermarket or frozen food locker plant.

When a local chain store has a "loss-leader" sale to attract customers, it is time for a freezer owner to buy up several dozen cans of frozen fruit juice or half a case of advertised-special frozen vegetables which the store manager is using for bait.

New frozen food variety items are frequently tested at low prices when they are introduced in order to feel out the market acceptance. The alert freezer owner buys them. They are bound to be good, for many months of painstaking research went into their production. It won't be long before the food wholesaler is satisfied that the price of his new specialty can be raised.

Learning to buy and repackage the frozen foods which are

sold in large institutional packs to hospitals, restaurants and hotels can save 25 per cent or more over retail prices.

How many times does a housewife throw away leftovers that the family cannot be made to eat? Perhaps it isn't much at a time, only a few cents' worth of vegetables or the ragged end of a roast. Also thrown away, however, is a proportion of the time it took to shop, prepare and cook the food, plus part of the gas or electric fuel consumed. If these few cents are multiplied by the number of days they are wasted throughout the year, the decimal point moves over two spaces.

A fisherman or hunter in the family, or even outside the family, can put many dollars' worth of good main courses in the freezer.

A friend in the country with property overgrown with berry bushes whose fruit is usually left for the birds will be more than happy to let a freezer family gather pailfuls in return for a few frozen packages after the berry season.

Why should a busy woman bake one pie or cake when the same amount of fuel and very little extra effort can put six or a dozen in the freezer?

And one of the intangible benefits of freezer ownership is that the better food consumed is reflected in family health. Research has demonstrated that freezer families eat more meat and fresher vegetables than the average for their income levels. What about the money saved on doctor bills, cold remedies and the like?

I don't pretend that the initial cost of a freezer will be returned during the first year of ownership, perhaps not for several years; but I don't know of any other single large-ticket purchase which pays off so dramatically and so consistently.

The chief concern of this book is to acquaint prospective,

new and veteran freezer owners with the latest tested and proved ways to live more conveniently and securely while enjoying better meals. The procedures are singularly simple; the advantages and economies are multiple.

H. M.

The Complete Book of HOME FREEZING

1. How It All Began

Has it ever occurred to you to wonder just how it came about that the many foods we eat, and the way we eat them, were discovered? What young Indian brave or squaw, for example, first looked at a tall, fibrous plant and observed the jutting, silk-fronded ears and was curious about how they would taste? Did he eat them raw and whole, or was he directed by a mysterious instinct which led him to discard the outer leaves and bite into the creamy kernels?

It may have been that a wanderer on a long-ago beach watched sea-birds swoop down at low tide to engage in battle with scuttling crabs, pecking at them with sharp bills until the crusty creatures were motionless. Our wanderer may have watched with silent interest while more birds joined the party and had a noisy, quarrelsome feast.

What intrepid soul ate the first oyster? The first mushroom? Who, digging in the earth, turned up a soiled, stone-like object and thoughtfully (or carelessly) bit into it, thus becoming the unrecorded pioneer consumer of a potato?

Ever since the world began, all living creatures have sought food in their struggle for survival, but the human species has developed a complex system of refinements which make eating pleasurable as well as necessary. These refinements include studied cultivation of crops and livestock, cooking ingenuity and artistry and, of course, the preservation of fresh food supplies for later use.

The early Egyptians are credited with "inventing" bread, for they discovered that the edible seeds of certain tall grasses could, when ground between stones and mixed with liquid and allowed to ferment, be fashioned into loaves and baked in crude ovens to yield a satisfying form of food. Moreover, they taught themselves to put aside some of the harvested grass seed and return it to the earth, where it would grow again the following year.

This was a tremendously important step in the progress of Egyptian civilization, for it meant that tribes need not travel constantly to far places in order to seek out new stands of the life-sustaining grasses; they were able to stay in one spot and grow their own. They could forsake the ways of the nomad and become land-cultivating householders. They could even grow more grass than they needed at one time and store it away for the future seasons, while the earth was giving birth to the next year's supply.

This was one form of food preservation.

Another early means of food preservation, picturesque to contemplate but not exactly in keeping with our modern sanitary sensibilities, was employed by the armies of ancient kings in marauding search of conquest.

The savage warriors thrived on a diet of huge slabs and chunks of fresh meat. Skillful hunters were sent out from the camping place to track down and kill ranging animals of the countryside, which were thrown into the campfire and later torn apart and eaten at the site.

When the warriors traveled on to new battles, however, they had to leave the uneaten portions of the carcasses behind them and take their chances on finding other animals the next night. This was irksome and occasionally hazardous, for the sounds of battle often frightened the animals into hiding. On such nights the tired soldiers went hungry, which did not put them in a mood for the next day's furious ride into combat.

One of the warriors, probably a remote ancestor of the world's first home economist, devised a way to solve the diet problem. Instead of leaving the previous night's unconsumed carcass behind as a grudging gift to carnivorous birds, he cleaved it into slices the size of giant steaks and instructed the warriors to use them as saddles on their bareback horses.

The soldiers fought naked in those days, and the horses were ridden with merciless, foaming speed. The body salts of rider and horse not only seasoned the meat to primitive tastes, but also served to keep it from spoiling as rapidly as it would have without the salt treatment.

This form of meat preservation we know today, under somewhat more prophylactic conditions, as "corning" or "pickling in brine," two popular and widely practiced methods for protecting fresh meat from the destructive and putrefying attacks of bacteria.

As the centuries passed, mankind's search for improved ways and means of preserving precious perishable food supplies resulted in such further devices as drying (dehydration) either by the sun's rays or by artificially applied heat; smoking; fermenting, and canning.

But, while these methods prevent or retard deterioration and spoilage, they also invariably *change* the food to varying degrees. The taste and texture of dried or corned beef are inevitably and infinitely different from those of a juicy roast; dehydrated mushrooms can never again resemble the plump, smoothly rounded caps of their original state, and canned fruit has an entirely different flavor from fresh.

Although dried, salted or canned foods are good foods and certainly contribute to convenience and menu variety, the ultimate goal of food chemists, technicians and researchers has always been to discover and perfect a way of preserving perishable commodities for long periods of time while retaining the typical taste, texture and nutritive value of their original fresh quality.

Refrigeration, both with natural ice and by artificial means, pointed the way to this goal. Freezing achieved it.

It must not be thought that the freezing storage of foods is new to the history of the hungry world, even though home freezers as we now know them post-date television sets on the national merchandising scene. Nor is food freezing an invention of twentieth century scientific endeavor. For centuries past, Eskimo tribes as well as other inhabitants of arctic regions have preserved seal, caribou, walrus meat and fish in ice caves or buried in frigid shoreline sand.

It has been reported that the first American foods were frozen artificially in the United States by crude methods, and frozen meat was shipped to England as early as 1876. By the first decade of the present century many commercial firms were in the so-called frozen food business. Because of inadequate facilities, however, which were more properly "cold-storage" rather than freezing methods as we know them today, the quality of the earliest attempts to produce frozen foods was inferior, and the products were not looked on with much favor by consumers, retailers or government bureaus.

It remained for a fur trader and biologist, the late Clarence Birdseye, to blaze the way to successful food freezing.

Mr. Birdseye said that the beginning of his interest in quick-freezing was, to a certain extent, accidental. In 1917, after having spent several years in Labrador on a fur-buying mission coupled with duties as a field naturalist with the United States Biological Survey, he set about developing a method which would permit the removal of inedible waste from perishable foods at their production points.

He envisioned an efficient and sanitary way to deliver a wholly edible fresh fish product to housewives in convenient, compact containers. Such an accomplishment would, he reasoned, offer a far more pleasing prospect to home cooks and also cut down considerably on transportation charges.

Unaware of experimental freezing methods practiced at the time in Europe and the United States, the biologist was puzzled as to how dressed perishable foods could be handled by the many necessary trade channels and yet reach distant consumers in a truly fresh condition. He realized that his goal precluded the use of packers' prevailing methods of preservation such as salting, smoking, drying, pickling or canning, for his ambition was to provide housewives with fresh fish whose appearance, color, flavor and texture would faithfully approximate their original quality.

Mr. Birdseye's first attempt misfired when he undertook to dress fish at his plant and chill it thoroughly before packing and shipping it in heavily insulated corrugated fiberboard containers, also chilled. Despite his precautions, he soon discovered that the product was still too perishable to be safe or practicable when shipped to distant points. Time and temperature variations during transportation proved to be destructive enemies to the food's freshness.

Probing deeply for a method which would overcome the spoilage obstacle, Mr. Birdseye remembered the fine, good-as-fresh quality of some of his Labrador meals. He recalled that some of the game, fish and imported fresh vegetables served to him could not possibly have been bagged, caught, or harvested the day or even the week before. He remembered further that these foods had been frozen with utmost speed in the bitter arctic weather which reached temperatures far below the normal freezing point.

Here, the scientist reasoned, was a clue which warranted testing; and so he developed a simple apparatus for freezing cleaned and dressed fish in moisture-proof packages at the extremely low temperatures which sped the process.

While Mr. Birdseye was conducting his experiments and developing the process which still bears his name, food packers in other parts of the country were also toying with frozen foods. These latter, however, were in the spotlight of disrepute, for the processors themselves were not entirely sold on maintaining highest quality standards, possibly because the idea's newness implied low prices. They were induced, therefore, to select for freezing raw materials of inferior quality, thereby revolving in the orbit of a vicious cycle.

It was not long before frozen foods *per se* were frowned on by federal, state and municipal authorities who set up rigid regulations.

The Birdseye method, however, was almost immediately acceptable, first to food-control agencies and later to the public itself. The Postum Company, Inc., soon changing its name to General Foods Corp., purchased the biologist's patent and embarked on a wide-scale pioneering operation.

By the spring of 1929, packaged quick-frozen foods began to reach American households. Varieties of food items were limited at first in patient deference to the public's skepticism. Only 80,000 pounds were sold the first year to adventurous and open-minded consumers. Today increasing billions of pounds of frozen foods of all kinds—fruits, vegetables, poultry, meats, seafoods, prepared specialty dishes and juice concentrates—are consumed each year by the American public.

The quick-frozen food industry, which includes seedsmen and nurseries, farmers, ranchers, fishing fleets, packers, processors, brokers, distributors, transportation facilities, refrigeration companies, material suppliers, warehouse real estate and retailers, is one of the largest and fastest-growing businesses in the United States. Not too many years ago, it was nonexistent.

As with most creative and useful enterprises, one idea begets another. The popularity and prevalence of home freezing and the vast equipment and food companies operating today undoubtedly owe their very being to Clarence Birdseye's application of his Labrador observances to his struggling fish-packing business.

Frozen foods required to be shipped in refrigerated freight cars and trucks to maintain their fresh quality, and so engineers were put to work to produce the vehicles. Retailers soon realized the need for low-temperature refrigerated cabinets in which to store properly their incoming deliveries of frozen food packages until these were purchased by customers, and so large commercial firms with refrigeration experience created new departments to design, manufacture and supply such cabinets.

Before long, farmers and ranchers began to wonder if they could utilize the new freezing methods to preserve some of their produce for their own families' use. For years they had resented the necessity of selling their livestock and food products to processors at low prices, later buying them back at much greater cost.

In various farming areas, frozen food locker plants were constructed to meet this rural demand. A few smaller freezer cabinets began to make their appearance, modeled after those used in retail stores, for the same farmers and ranchers recognized the convenience of having additional freezer storage space close to the kitchen in which to keep a week's or a month's supply of poultry and meat brought home from the community locker, and to use for freezing smaller quantities of seasonal fruit, berry and vegetable crops.

The idea was catching on, and American business leaders rarely permit a new idea to remain lonesome. A few far-seeing manufacturers of refrigerators and other home appliances began tentatively to produce freezers for home use, although the general public was not yet ready to accept them.

When the United States entered World War II freezer cabinets, like so many other war-scarce products, were discontinued because they were made of materials vital to the war effort. But food was scarce, too, especially meat and meat products. Fresh and most canned food was rationed, but for some reason frozen foods were ration-free.

Out of desperation, many housewives who had previously ignored or scorned frozen foods began to buy them because of their availability. Surprised and delighted with their excellent taste and quality and the convenience of preparing them, women continued and repeated their purchases of the revolutionary foods.

At about this time, some people became aware of so-called "wholesale" food purchasing; for while steaks, chops, roasts and hamburger were rationed, sides or hindquarters of steers, calves, hogs and lambs were not. Those lucky enough to have freezer space either at home or in community locker plants ate better than their neighbors did.

Many enterprising families sought and found second-hand commercial low-temperature refrigerated cabinets, others snapped up used ice-cream freezer cabinets or milk-cooling cans and converted them, a few mechanically-talented husbands designed and constructed homemade freezers.

In kitchens and living rooms in all parts of the country, parents and visitors, not yet personally familiar with either freezers or frozen foods, listened with nodding approval and with mouth-watering intentness to descriptions of excellent meals many of their on-leave sons and daughters were receiving in the armed services. They learned that army, navy, air force and marine cooks were being supplied with frozen foods from the various quartermaster centers.

The idea continued to grow, attaching itself firmly to the minds and imaginations of American homemakers.

After the war years, during the limping period of conversion, manufacturers began again to produce home freezers. The output was small, at first, for so many other products were also in demand. Besides, the home freezer market was not ready-made. Long and costly educational advertising and publicity programs had to be evolved and sustained in order to let the people know the advantages, economies and conveniences of freezer ownership.

As acceptance grew, so did production. In an amazingly short period of time, home freezer manufacture and sales have far outdistanced even the most optimistic predictions made for them by business forecasters. It took manufacturers of mechanical refrigerators ten years to arrive at the number of sales reached by home freezers in four years.

There are excellent reasons why this is so.

A home freezer offers joys and advantages unmatched by any other single item of household equipment. It is, moreover, the symbol of a new and revolutionary phase in modern living. In this great age of mechanical devices which are deliberately designed to make life easier for everyone, the only phase of daily living which had lagged behind the times was the method by which housewives bought their food and prepared it for the table.

Women no longer haul water from an outside pump; they

take for granted gleaming sinks, hot and cold running water and efficient drains. Homes are no longer lighted by evil-smelling, smoky, hazardous kerosene lamps; a flick of a manicured finger floods a room with brilliant electric light. The corn broom and the back-breaking rug beater have almost passed into oblivion, replaced by vacuum cleaners that all but operate themselves. Old-fashioned laundry tubs gather cellar dust as washing machines swish away dirt from clothes and household linens. There are not many blacksmiths, these days, and the lineal descendants of buggy whip and carriage-accessory craftsmen must certainly have found some other less anachronistic business to go into by now.

But, until frozen foods and home food freezers came along, women still bought their food the way their grandmothers did—a little at a time, day after weary day, as doggedly and faithfully as a postman on his appointed rounds. Neither rain, nor sleet, nor aching feet prevented them from marching to the corner store or the neighborhood supermarket. They jostled elbows, scrabbled for bargains, tried to stretch the rapidly ebbing dollar to cover the more rapidly rising cost of food. Then they trudged back home again, arms laden with heavy bundles, to begin the chore of preparing the food for the table. Next day, or the day after, the whole tiresome routine had to be repeated.

This monotonous, time-consuming, frequently inconvenient repetition was, however, necessary for every conscientious wife and mother who took seriously her responsibility for her family's well-being.

Food, according to all she had been taught, was best when fresh from the market and tastefully prepared. In addition, she knew her husband's and children's preferences in meals and tried to please them with a variety of interesting menus throughout the week. In order to do this, she had to shop constantly and continuously and plan the pattern and balance of her meals with the strategic canniness of a five-star general.

She probably once tried feeding her family the Sunday roast in other guises on Monday, Tuesday and Wednesday. She probably tried it no more than once, however. It is legendary that leftovers are usually greeted by the family with a rude groan, an outraged bellow or a cold, injured silence.

The first glimmers of grand new vistas ahead reached the housewife in the form of increasingly improved commercially packaged fresh-frozen foods.

America's homemakers welcomed frozen foods with open arms and grateful hearts. Frozen foods are attractive. They are easy to buy and easy to prepare. They take much of the guesswork out of marketing, and almost all of the drudgery out of preparation. They are fresh. They are nutritious. They usually come packaged along with careful instructions and, in many instances, with idea-inspiring recipes.

Incidentally, frozen foods offer one subtle but significant feature which I have never seen used as an advertising angle: They make obsolete the snide little jokes about a bride's biscuits or a wife with a can-opener. Few husbands or even mothers-in-law have sarcastic or disparaging comment to offer about an enticing meal prepared with fresh food from the freezer.

The frozen food processors have spent many years and many millions of dollars in research and experiments in order to make their products appealing to the public. They have learned the right way to grow, select, freeze, package and deliver perishable foods so that when they reach your supermarket's freezer cabinets they are still safely in the frozen state, ready to be prepared for your dining room table.

Although the main purpose of this book is to assist you in

the purchasing, planning, preparation and packaging of fresh foods for your home freezer and to suggest various ways of cooking them after they are removed from frozen storage, there will be many times when your best interests will be served by buying a dozen packages or a case of a commercially packed frozen food item.

Because you have a home freezer you will be able to take real advantage of competitive "specials" from your local stores. The more you buy, the more you will save. In addition, many food retailers offer case-lot discounts to home freezer owners as a general practice.

But your home freezer has many, many practical, convenient and economical uses other than as a storage place for store-bought frozen food packages. These are the uses with which we will concern ourselves throughout the pages of this book.

2. Freezers in General

At the latest count, there are upwards of seventy-five reputable firms in the country who manufacture home and farm freezers ranging in capacity from two-cubic-foot "portable" models to room-size "walk-in" types approaching 250 cubic feet or more.

There are square chests, oblong chests, squat uprights and tall uprights, some with fast-freezing shelf plates and some without. There are still others with inner doors and pull-out drawers.

There are white porcelain and enamel exteriors, stainless steel, aluminum or bright copper exteriors. There are pastel-tinted finishes to match the new color trend in kitchens, and there are cabinets which combine a full-size freezer with a full-size refrigerator. There are built-in models and hanging models.

With this confusing wealth of selection available, it is

wise to check-list individual needs before making a decision about size, style and brand.

There is a prevailing phenomenon, known to psychologists and sales analysts, which prompts a prospective purchaser to do a little private sleuthing among friends and relatives who have already taken the plunge and bought the product under consideration. This is such a universal practice that it has become traditional.

If you have not yet bought your freezer, however, I should like to warn you away from that tradition. It is human nature to endow personal possessions with superlative qualities. Every mother's child is unusual. Your freezer-owning neighbor or sister-in-law, commendably eager to share with you the freezer advantages they so happily enjoy, are more than likely to say, "I have a Stratosphere Freezer, and it's out of this world! If you're going to buy one, don't even consider anything except a Stratosphere."

This is a charming manifestation of loyalty, and the majority of American business leaders have learned to lean heavily on its prevalence. In hushed and reverent tones around sales and advertising conference tables, they call it "brand consciousness and consumer loyalty." It is very nice, but it is not always conclusive.

Your freezer-happy neighbor, for example, probably has needs and problems which are not identical with yours. Her Stratosphere, possibly a handsomely designed nine-cubic-foot chest with a built-in fast-freeze compartment, fits beautifully in her kitchen floor plan, right next to the refrigerator and several paces away from the stove. She has a husband, a son away at college and an adolescent daughter who is on a determined diet. She entertains occasionally, and, so far, she has found her freezer very adequate indeed for her needs.

You, on the other hand, may have a larger family, a smaller kitchen and a far busier schedule. Whatever your circumstances, it is more than probable that your requirements for a freezer are going to be different from those of anyone else. For this reason, do resist the temptation to buy a twin to someone else's freezer—at least until you measure your own needs with a prophetic yardstick.

First Considerations

Although a freezer may be used as a sort of luxurious gadget in which to keep on hand a supply of out-of-season delicacies, party refreshments and the limit of pheasant your husband bagged a few months ago, it is not the general rule. You would hardly buy an automobile to use exclusively for transportation to the theatre.

A freezer is a hard-working, utilitarian, cooperative partner in everyday living. It is an investment that will pay off in dividends many times more than its original cost. Approach the purchase of a freezer in much the same spirit with which you approach the purchase of a home or a family car. The idea is not to get the cheapest one available, but the one which will give you the most value in the long run, whatever the initial price tag.

Keep in mind that most sources which sell home freezers offer financing terms within governmental limitations, and will tailor a convenient time-payment plan to your individual needs. Those lucky enough to be able to pay cash may, of course, save the financing charges and receive an additional cash discount as well.

Be very sure, however, that the firm from which you purchase your freezer stands ready to service it whether you pay cash or buy on time.

Fit the Freezer to Your Family

In most cases, the freezer will be used to contain and supply as much as possible of all the freezable perishable foods consumed by your family. One of the frozen food industry's trade magazines states that 75 per cent of all the food we eat is perishable, and goes further to predict that eventually 75 per cent of all the food we eat will be prepared for the dinner table from the frozen state.

If the master of ceremonies on a TV quiz show were to ask you to estimate in pounds how much food you eat during a year, you would probably answer, after brief and rapid calculation, "Oh—about six or seven hundred pounds, I guess." You would be wrong. You eat more than 2,000 pounds of food a year. Does this sound fantastic? It is actually a conservative figure.

The United States Department of Agriculture and the United States Bureau of Human Nutrition have broken down the annual per capita consumption of all the foods eaten by all persons residing in this country. One list itemizes everything from turkey (5.6 pounds per year per person) to peanuts (4.4 pounds) and includes such items as meat (138 pounds), fresh vegetables (252 pounds), fresh fruits (119.1 pounds) and wheat (in breads, cakes, cereals, etc.—133 pounds). These are generalizations, of course. You may not eat a peanut all year, but, on the other hand, you may eat 10 pounds of turkey.

Altogether, the various foods on the government list add up to more than 2,100 pounds per person. This, however, is basic dry weight. For example, the average American adult consumes 16.4 pounds of dry coffee per year. The weight of the water is not included. Some of the 252 pounds of fresh vegetables are consumed in the form of soup, yet only the weight of the vegetables themselves is computed.

While it is true that the water content of soup or stew cannot properly be included in the weight of actual food consumed, remember that here we are discussing freezer capacity for your family. When you freeze soup or stew, you freeze the water along with it, and it takes up freezer space.

It has become a cliché of home freezer advertising to say that one cubic foot of freezer space will hold approximately 35 pounds of food. It is well to consider, however, that while one cubic foot of freezer space will geometrically hold 35 pounds of solids, the food you store in your freezer will not usually be geometrically shaped in order to fill up every corner and crevice of the cabinet.

Many freezer packages such as meats, poultry and fish are uneven in shape and size. Foods like cake and cauliflower are light in weight, but bulky. It is closer to reality, therefore, to pare a few pounds from advertising copy which says, "Ten cubic feet of freezer space—holding 350 pounds of food."

If you plan to put nothing but square packages in your freezer, count about 32 pounds to the cubic foot, for the packaging materials themselves take up some room and so do the air spaces left for necessary expansion in many fruit, vegetable and pre-cooked food packages.

If, like most people, you plan to put a miscellany of variously packaged foods in your freezer, it is better not to count on more than 25 or at most 30 pounds per cubic foot.

Government bulletins and many home economists state that six cubic feet of freezer space should be allowed *per person* if the freezer is to be used to store a major portion of the family's perishable food. Many manufacturers' ads, I suspect in order not to frighten you, say that three cubic feet per person will suffice.

Personally, I agree with the government. Three cubic feet of freezer space per person may be enough for a family which also uses the freezing and storage facilities of a community frozen food locker plant, but they simply are not enough for a family which takes full advantage of freezer benefits by storing the bulk of its fresh food conveniently at home.

Six cubic feet will hold approximately 180 pounds of food, and wise freezer owners will "turn over" their freezer contents three or four times a year. This means that each person, with six allotted cubic feet of freezer space, can expect an annual yield of from 540 to 720 pounds of food from the freezer. That is not much, considering that he actually consumes a total in excess of 2,000 pounds throughout the year.

So, What Size Freezer?

This somewhat mathematical discussion has had one purpose: To lead you gently and without undue alarm to the conclusion that a 20- or 21-cubic-foot freezer is not enormous if you are the average American family of 3.72 adults. Incidentally, the word "adult" in this context is disarming, for it applies to the quantitative eating ability of the individual. Have you fed many growing boys and girls? Legally, and even to the statisticians who gave us the 3.72 figure which represents the American family average, growing youngsters are not classified as adults. As a measurement of gastronomical capacity, however, they may certainly be counted as such.

In other words, a family of four *needs* a 21-cubic foot freezer.

Will you, your husband and your 1.72 children be happy with a 6- 9- or 12-cubic-foot freezer? You will, but you will soon wish that it were larger. The more space you have, the more food in a greater variety will be on hand. The more food and the greater variety on hand, the more money saved on food bills, less time spent in shopping, freezing and cooking.

The best recommendation I can give you for choosing the size of freezer most suitable for your needs is to multiply the number of people in your family by five to arrive at the cubic footage to keep in mind when you start shopping for your home freezer.

If you have lots of room in your home, add a few more cubic feet for good measure. Don't be alarmed at the result. A freezer whose capacity is 21, 18 or 16 cubic feet may seem monstrous to contemplate, but I assure you that it is no such thing.

Many freezer-experienced families find it necessary, convenient and economical (after the initial cost) to own two freezers—an upright and a chest. They quick-freeze on the upright shelves for transfer to the chest, storing also in the upright when supplies warrant. In times of low reserves, they rest one of the freezers and live out of the other while they wait for seasonal buys or local harvests.

What Style Freezer?

Ordinarily, your choice of a freeezr style will be made between a trunk-opening chest and a door-opening upright. I would be less than scrupulously honest if I did not admit at once my own preference for the upright models, but I realize that many of the women who read this book have already acquired the horizontal chest style of freezer and may give me an argument.

I have owned both types and have used both with ease, joy and benefit. I happen to prefer the upright because of its floor space economy and because it is easier for me to reach in than to reach down. Moreover, the freezing plate shelves of my model, nestling as they do on refrigerant-containing coils, permit me to freeze foods with a maximum of speed and simplicity and then "file" them for easy accessibility.

As a new freezer owner learns very soon, it is as important to get foods out of the cabinet as it is to get them in there in the first place.

I would like to scotch a rumor which is abroad in the land to the effect that upright freezers, because of their outward-swinging doors, permit cold to "escape" and thereby reduce freezing efficiency. Let it be said here that this is the sheerest nonsense.

It is true that some cold air will rush out to greet the warmer air of a room when the freezer door is open. You can sometimes even see it as a vaporous cloud spilling Niagara-like over the frosty shelves. The actual loss of cold within the freezer, however, is negligible to the point of nonexistence. Except when a housewife is engaged in the process of putting newly packaged foods into the freezer to be frozen or in taking out a day's supply to be cooked, the door is rarely opened more than a few times a day and then only for seconds at a time. Not enough cold will get away to chill an already cold stringbean.

The interior of a well-constructed upright freezer is engineered to maintain a temperature of 0°F. or below and will adjust itself to that degree no matter how many times the door is opened during normal activities. If you have used packaged frozen foods bought from a store, you know that a 12-ounce package of frozen fruit or vegetables takes a long time to thaw, even in a warm room. Do not believe anyone who tells you that an upright freezer "spills out the cold" to any degree that is harmful to its contents or profligate of electric current. If the freezer is kept well filled, as it should be, the cost of cooling warm air is minute. Make your choice between the two styles, chests and uprights, for other and more realistic reasons.

What to Look for in Chest Style Freezers

EXTERIOR FEATURES

If you have plenty of floor space in the room where you plan to keep your freezer, you may decide on a chest style.

Well-made chests are available in many sizes, from four cubic feet upwards.

Be sure, before you sign a contract, that the width, height and length of the chest you have chosen are compatible with the measurements of the doorways leading to the room in which you plan to keep your freezer. The average interior doorway is seldom more than 30 inches wide, and some may be even narrower. Most chest freezers range in width from 27 to 33 inches.

Measure carefully the doorways through which your freezer must be carried, being sure to take the measurements between the inside moldings of the door frames. There have been many instances of last-minute jimmying off of door moldings to accommodate the passage of a freezer just a fraction of an inch too wide to get through an incorrectly measured opening.

When you consider the height of a freezer you are thinking of purchasing, consider also the height and arm-reach of the person or persons who will be doing most of the freezing and who will on occasion find it necessary to reach into the chest for food packages which may be tucked into far-away corners at the bottom.

While most manufacturers have standardized chests to table-top height, or 36 inches, to the level of stoves and kitchen cabinets, there are several brands on the market which are a few inches lower and some which are as high as 37 or 38 inches. A small, short-armed woman will have a hard time bending over to reach out-of-the-way packages without the aid of a step-stool.

The length of the chest style freezer you choose will depend on its cubic capacity. The larger the capacity, the longer the chest. Here, too, doorway measurements are important; for while sometimes an extra-long chest will clear the sides of the door in good order, it cannot be turned in either direction (in a small room, for example) for proper placing and may have to be up-ended in the doorway.

If you decide that the basement is going to be the place where you will keep your freezer, be sure that all dimensions will clear the space between the cellar steps and their wall and ceiling boundaries, as well as any turns.

With the increasing popularity of home freezers, it is conceivably possible that some time in the not-so-remote future will find architects reverting to the quaint custom of including "coffin niches" in their plans, renamed "freezer niches," of course. Many old houses to be found throughout the country still have these recessed niches, usually arched at top, located at the bend of narrow stairways.

Outer shell: The better chest style freezers are constructed of one- or two-piece heavy gauge steel, welded and braced to insure absolute rigidity. Some steel exteriors are permitted to remain unfinished except for treatment against rust or stain, but the majority of manufacturers have found that their women customers prefer the smooth, gleaming white or decorator finish they have become accustomed to seeing on refrigerators, stoves and other kitchen appliances.

To be sure of long wear and attractiveness, make certain that the porcelain-like enamel finish of the cabinet you choose has been *baked on* over a metal shell which has been pre-treated for rust resistance.

Remember also to check the very bottom of the chest for a few inches of toe-space allowance. Toe space is desirable, for it prevents scuffing, makes floor mopping easier and provides greater convenience in use.

Hardware: By "hardware" is meant any outside closure or lock, lid hinges and exterior metal trim.

It is extremely important that the hardware on a home

freezer be more than merely handsome; it must be exceptionally durable and rugged in order to withstand hard use under temperature-variant conditions. Rust resistance is a must for freezer hardware, because the outside of the cabinet may sweat.

Top-opening lid: For safety and ease of operation, a chest freezer's lid should by all means be counterbalanced on a hinge or balancing device which should be recessed in order not to interfere with food storage space within the cabinet. Once raised, the lid should stay in its upright position until pulled down by the user and, preferably, should be double-balanced at a point a few inches above contact with the cabinet to safeguard against the possibility of bruised fingers if the lid should be pulled down hastily or carelessly.

The operation of a properly balanced lid is much the same as the one with which you are familiar in the luggage compartment of an automobile.

Lid seal: Make sure the seal is airtight when the cabinet is closed. Most lids are fitted with a rubber or rubber-like plastic gasket strip to absorb shocks and insure airtightness. This is an important feature, for if air is permitted to pass through the cabinet's closing seal, moisture in the air will speed the collection of frost within the freezer and may also form a film of ice on the gasket, which will cause the lid to stick.

INSIDE THE FREEZER

Interior finishes on what is called the "liner" of the freezer are most usually either acid- and chip-resistant porcelain enamel or equally stainless steel or aluminum. The liner, like the shell, should preferably be constructed of one-piece welded steel (or heavy aluminum) and all joints and seams should be thoroughly sealed to protect against moisture infiltration and loss of insulating efficiency.

Some manufacturers of chest style freezers provide interior lights which illuminate only when the lid is raised, darken when it is lowered. This is an important feature to look for if you plan to place your freezer in a fairly dark corner, or in the basement.

Some freezers incorporate in their design an interior or exterior thermometer, which enables you to check at a glance on the temperature within the storage compartment.

Most chest freezers larger than four or six cubic feet in capacity contain sectional dividers, either permanent or removable, for greater ease in compartmentalizing stored frozen foods. Also provided are one or more baskets made from a non-corrosive material which are fitted across the top of the cabinet in such a way as not to interfere with the closure.

Many freezers come equipped with either an audible or visible alarm signal, whose purpose is to warn you in the event of power failure or any sudden mechanical disturbance which means that electric current has been cut off, endangering the frozen contents. If the freezer of your choice is not already equipped with such an alarm, one may be purchased separately and attached to the mechanism by an electrician.

Needless to say, the audible type—a bell or buzzer signal—is preferable if the freezer is placed in the basement or in a room which is not occupied normally. The visible signal, usually a winking red light, is adequate warning when the freezer is placed in the kitchen or a frequently trafficked utility room.

Before we embark on a comprehensive discussion of mechanical operating features of a freezer, which apply to both the chest and upright styles, let us examine briefly some of the comparable qualities to look for in the upright model, if this one is to be your choice.

What to Look for in Upright Style Freezers

EXTERIOR FEATURES

Generally speaking, the exterior dimensions of an upright freezer do not require as much careful checking with household space limitations as does a chest. Because of its skyscraper design, an upright usually takes up little more than the floor space demanded by a refrigerator. Inasmuch as most rooms have more available wall space than unused floor space, you need only compare the measurements of all doors through which the freezer must pass with the narrowest dimension of the upright freezer—usually its front-to-back depth.

Something to be considered when you order your upright freezer is the position it will occupy in the room where it is placed. In standard manufacture, an upright freezer door is right-handed, swinging outward on hinges concealed at the right side of the cabinet as you face it. The handle is located at the left.

If the only spot on your floor plan which will accommodate the freezer is hemmed in at the right with cabinets, immovable objects or a wall, you are likely to find a right-handed door too inconvenient to be practicable. Under such circumstances, ask your dealer to determine whether the upright freezer you have selected can be obtained with a left-handed door.

Rearranging the concealed hardware and the handle and re-hanging the door from the left side can be done at the factory. You may have to wait for delivery and pay a nominal charge, but this is better than constantly banging the door against an obstruction and not getting complete exposure of the freezer's interior.

Outer shell: Good upright freezers are constructed of one- or two-piece heavy gauge steel on a welded and braced metal frame. As with chests, the exterior is usually finished with a white or pastel baked-on porcelainized enamel, although a few manufacturers produce freezers with a stainless steel or copper finish. If porcelainized, the metal shell should be pre-treated for rust resistance.

Hardware: Hinges, handle and metal trim should be made of rugged, rust-resistant materials.

Door: An absolute airtight seal is essential to an upright door to prevent excessive frosting within the freezer.

A simple and fascinating way to test whether or not the door seal is perfect was told to me by a freezer engineer. Take a dollar bill and close the freezer door on it, leaving enough of the bill protruding on the outside to grasp with your fingers. Tug gently on the bill. If it can be pulled out, the door is not properly sealed.

In most upright freezers, it is a simple matter to adjust a door seal which may have been jarred loose in transportation. The men who deliver and install your freezer can do it on the spot. If the door works loose after the freezer has been in your home for any length of time, call the dealer from whom you bought it. He will either send a service man to adjust it, or will tell you over the telephone how you can do it yourself.

Other upright features: Interior lights and alarm signals, as well as thermometers, are supplied with some freezer brands. Certain manufacturers, catering to the unnecessary fear that "cold air spills out" of upright freezers, have equipped each freezer shelf area with a springed door, while others arrange the shelves as drawers which slide out to expose the contents, like a bureau.

INSIDE THE FREEZER

Although some manufacturers continue the porcelainized enamel finish on the interior surfaces of the upright freezer, my personal preference is for stainless steel or heavy gauge stainless aluminum interior walls and shelves. These serve additionally to speed freezing efficiency, for they are generally better conductors of cold than are surfaces coated with plastic or enamel. They have another advantage, too. Most defrosting of upright freezer shelves (except for total defrosting, described in the next chapter) is done with a blunt scraping instrument; metal shelves resist scuffing and scratching better than do porcelain ones.

Mechanical Features of Both Chests and Uprights

These, invisible to the unpracticed eye of a housewife who is not wise in the ways of technical engineering, are vastly more important than a handsome appearance. Not every brand of freezer offered to the public has all of the exterior and interior characteristics we have discussed; whether or not all of the accessory features are necessary for freezer felicity is open for debate, and in the final analysis must probably depend on individual preference.

When it comes to intrinsic mechanical features, however, particular care should be exercised in pre-purchase shopping and comparison in order to insure maximum quality and dependability of performance. After all, it is for performance that you are buying a freezer, not for looks.

It is usually a good idea to enlist the interest of your husband or another construction-minded man in helping you to make your decision. As manufacturers of everything from automobiles to roofing material have long since learned, a smart appearance is an effective way to woo a woman's purchasing power. Too frequently, the ladies are likely to overlook fundamental features in favor of pretty ones. A common dialog between husband and wife looking at a house they are considering usually goes something like this:

Wife: What a perfectly darling kitchen! Look, it has a picture window right over the sink. I can watch the children while I do the dishes.

Husband: H'm. Copper plumbing. That's good. But I wonder if that window has permanent attachments for storms and screens?

Wife: Don't you love the way this house hugs the ground? What a setting for really smart landscaping!

Husband: Foundation isn't deep enough by at least two feet. Be damp in bad weather.

And so it goes, with variations. A woman is primarily concerned with convenience and charm, a man with basic construction features.

Unlike a home, however, a freezer's important operating efficiency cannot be determined beforehand without taking the appliance apart. As a rule, you must rely on the integrity and reputation of the manufacturer, who states his product's construction features on the specification sheets supplied to customers through accredited dealers. *Insist on studying these specification sheets* (called "specs," in trade jargon) for every freezer you seriously consider. Insist, too, on being shown exactly what *printed warranties* you will receive with each brand. They vary.

Assisted by structural features such as positive door-seal and the sealing of all exterior joints and seams against moisture infiltration, the principle which makes your freezer *freeze* is a skillfully constructed refrigerating mechanism called the condensing unit.

This system is composed of a motor-driven compressor, condenser, expansion device and evaporator and its function is to circulate the *refrigerant* sealed into the coiling system.

A fully automatic control monitor starts and stops the compressor when the temperature within the freezer varies from the "set" degree. This can be compared with the action of a thermostatic control on your heating system, which seems to know when the temperature in your house goes below 65 degrees and sparks your furnace into operation until the desired temperature is reached, at which point it automatically shuts off.

INSULATION

Needless to say, the most perfect operation of these mechanical principles would be wasted if the freezer itself were not properly constructed to *retain* the cold produced, any more than you could keep your house warm by running your furnace at high heat if all the windows and doors were left open to allow the cold to enter.

Retaining the cold within your structurally airtight freezer is the job of the insulation packed between the outer and inner shells of the cabinet and of the lid or door. This is almost invariably a specially packed fiberglas material in batting form, varying in width from 1½ to 6 inches. A few models, especially uprights, are being insulated with a slimmer, high-density material which permits manufacturers to streamline their designs by cutting inches off the all-over exterior dimensions.

THE CONDENSING UNIT

This is the heart, soul and spirit of the refrigerating mechanism in your freezer. When you read the specification sheet provided by the manufacturer of the freezer you are examining, you may not find this word listed. Some manufacturers identify each separate part of the mechanical system by name, others gather all parts under the general classification of "power system," "mechanism" or "unit," still others mention one or two parts and omit the rest.

Most manufacturers do name and qualify both the type of refrigerant used, and the insulation material. Whichever way it is identified, however, the condensing unit should be described as *hermetically sealed.*

Refrigeration engineers have found that the hermetically sealed multiple unit is more desirable for home use than the open type, which is usually placed at a remote distance from the freezer itself. The sealed multiple unit is compact and quiet in operation. It never requires oiling, as a lifetime oil supply is sealed into the housing. In such a sealed mechanism, moreover, there is no danger of refrigerant leakage.

POWER

Most home freezers of either chest or upright style operate on household currents of 110 or 115 volts, 50 or 60 cycle, single phase, A.C. If you are uncertain as to the kind of power you have, it is wise to telephone your local electric company for correct information. If you live in one of the few remaining localities in the country which receive direct current (D.C.) it will be necessary for you to purchase a transformer.

If possible, the circuit you are going to use for your freezer should be provided with a separate fuse of proper amperage. It is not a good idea to use the same outlet for any other appliances, as too great a load may cause a short circuit.

Motors of ⅛ horsepower and up are used for home freezers, depending upon their size and design. The higher the horsepower, the colder the freezer will be, all other factors such as cubic area, airtightness and insulation being equal. This is to say, a 10-cubic-foot freezer with a ¼-horsepower motor will maintain lower temperatures than the same size freezer with a ⅕-horsepower motor.

FREEZING SURFACES

In the simplest type of chest freezers, the coiling containing the refrigerant which effects the freezing is usually sealed into the side walls of the cabinet, and occasionally also to the underneath side of the cabinet's floor, being located directly in back of the liner. This type of chest cabinet is called a "wrap-around."

For some chests, particularly those of larger storage capacity, the manufacturers have provided separate compartments for sharp freezing. In these models, the cabinet has two different temperatures, the sharp-freezing section as low as 10 or 20 degrees below zero, the larger storage section maintained at a constant of zero or at most 5 degrees above zero. All temperatures are given in the Fahrenheit scale.

In upright freezers, it might be well here to distinguish between those which are essentially just wrap-around chests turned on their sides, with shelves added, and those with *freezer plate shelves.*

The distinction is more than just superficially important, and if you want the fast-freezing advantages of owning an upright rather than just the convenience features of greater accessibility of packages and the elimination of bending over and reaching into the cabinet, you had better make certain that the upright you are considering has freezer plate shelves.

In wrap-around chests and uprights alike, the freezing of food packages is done by placing them in contact with the side walls; in freezer-plate shelf uprights, however, the freezing is done by refrigerant-filled coils located directly beneath the shelves, usually three or four in number. In such uprights, the law of gravity sees to it that your food packages maintain positive contact with the freezing surfaces when they are placed on the shelves. The distance between one shelf and the next rarely exceeds 14 or 15 inches. This means that even after food packages are frozen and stacked to make room for more packages on the contact shelves, they are never more than just a few inches away from intensely refrigerated surfaces.

The bottom shelf, or floor, of most uprights is also provided with coils, but the heat of the condensing unit may raise the temperature of this section a few degrees. It is generally better to use the bottom shelf of an upright for storage only, doing the fast-freezing of food on the three or more freezer plate shelves under which you can actually see the coils.

To get the best performance from an upright freezer, therefore, be certain that the shelves are freezer plates, each with its own system of coiling.

A Special Word About Manufacturers' Warranties

When you embark on your freezer-comparison expedition, train yourself to read guarantees very carefully. Reputable manufacturers are explicit as to the conditions covered in their printed guarantees, and the time period to which the guarantee applies is clearly stated.

When you run across big, black capital letters which proclaim "GUARANTEED FOREVER!" or "LIFETIME GUARANTEE" read the small print to ascertain just what it is that is so impressively pledged for all eternity. Do you remember those fantastic life insurance policies which blazoned, "$10,000 BENEFIT PAID AT ONCE TO SURVIVOR!"? In very fine print, buried in the text, came the explanation: "If insured is gored by a charging bull or struck by lightning."

Some freezer manufacturers do guarantee the satisfactory operation of the mechanism for as long as five years, and the cabinet itself for one year. A few firms also offer as much as five-year protection against food spoilage, promising to reimburse the purchaser in cash for actual loss of food by spoilage due to mechanical failure. This additional protection, however, is not determined by the manufacturer's good faith alone; several individual states have enacted legislation which prohibits such warranty, no matter how willing the manufacturer may be to offer it. It might be a good idea to ask your insurance broker about the law in your state before you believe a salesman's assurance that you will be protected against food spoilage.

Be sure to ask your dealer for all printed guarantees when you conclude your purchase, and keep them always in a readily accessible place. You will ordinarily be asked to sign a registration card on which will appear your name, address, date of purchase and the serial number of the freezer installed. This is for your own protection.

Because most freezers are scrupulously checked during manufacture and thoroughly tested before being shipped, you will probably never need to enforce the guarantees in your possession. It is wise to insist on them, however, not only "in case," but also because they are evidence that the manufacturer is proud of his product and is willing to assume responsibility for its satisfactory operation.

From Whom Should You Buy Your Freezer?

An easy generalization would be—buy your freezer from an accredited electric appliance dealer in your community whom you know and trust. Like most generalizations, however, this one also is too broad to be conclusive.

In many sections of this country there isn't an accredited electric appliance dealer within miles; or, if there is one, he may stock only one or two brands. You should not have to be bound by such limitations. Every freezer manufacturer will be happy to send you specification sheets and descriptive literature about his product. If the brand of your choice does not have a dealer in your immediate vicinity, the manufacturer will be more than eager to send you the name of the nearest one.

It is, moreover, perfectly safe to order a freezer from a reputable mail-order house, or via mail from a large-city department or appliance store which makes deliveries in your locality. Just satisfy yourself first that the manufacturer's reputation, advertised descriptions and warranties are trustworthy and that the retail source will assume responsibility for delivering the unit as represented. You do not buy an unknown quantity when you buy an honest manufacturer's advertised brand from an honest dealer's showroom or warehouse.

What About "Food Plan" Purchases of Freezers?

Time was, and maybe still is, when a local motion picture theatre would advertise "Dish Night." This was a night when every person going to the movies to see one or more films would be given at the door a saucer, a cup or a dinner plate in an attractive design. If you did not move out of the neighborhood and if you went to the movies regularly on "Dish Night" it was possible, especially if you were blessed with a large family, to accumulate a whole set of dishes free. This custom, sometimes called "building good will and establishing brand consciousness with premiums," is also practiced in various forms by many manufacturers wishing to attract new customers. A soap firm will offer plated silver spoons in return for coupons wrapped around large, economy-size bars of soap. Breakfast cereal companies are wily in their ways; they get children to insist that their mothers buy Beentzies because Beentzie box-tops can be used as wampum in exchange for model nuclear fission laboratories or other exciting toys. Television and radio pitch-men plead with you to accept—absolutely free—a 375-piece set of kitchen cutlery, yours for only a telephone call and $1.00 a week toward the purchase of an all-electric outside barbecue pit.

The premium, or something-for-nothing, business is a large and thriving one and no one is very startled to be offered a related or unrelated "free gift" as an inducement to purchase something else.

Within the last few years, however, the housewives of America have been somewhat pleasantly surprised to find in many areas of the country that the premium they are being offered is *food,* which represents the largest expenditure of a family's annual budget.

Fresh, high-quality food, the enticing advertisements state in newspapers or over radio and television, will either be given to you without charge or sold to you at ridiculously low prices. Of course, you can't possibly keep all that food around the house without something sensible to put it in. You'll need a freezer, naturally. But all you have to do to enjoy low-cost food is buy the freezer on easy payment terms, spaced out over a period of time up to three years.

The proposition is exceedingly attractive and many householders have rushed into adoption of such plans without stopping to analyze them to arrive at a proper evaluation.

Despite the howls of protest that have arisen from many business sources, the so-called "freezer-food-plan" is basically a sound idea, offering the formerly freezerless consumer new conveniences, leisure and economy. The indignant protests, for the most part, come from retail food markets who fear (with some justification) that the food plan will take away their customers; from large frozen food interests who are reluctant to endorse anything which might antagonize *their* customers, those same retail food markets, and from regional freezer dealers or distributors who are not involved in a food-plan set-up but who see prospective sales being clinched joyously by those who are.

These dealers and distributors cry "unfair practice" until they themselves come to the point of sponsoring a food plan; then they have another word for it—"competition."

I have been intimately associated with the freezer and frozen food industries for many years, and have worked closely with one of the country's foremost food-plan originators. I have this to say: Any method which puts any good, reputable freezer in a home at its list price or below and which stocks it with good, nourishing food at standard or reduced prices and which provides the means for financing the whole package out of income—that's for me. It's for you, too, if you approach the idea cautiously. The merchandising method may be unorthodox, but it certainly gives the customer a break.

The food plan has, to my mind, a secondary virtue as well. The time has not yet arrived when people in general regard freezers or freezing with the casual acceptance they deserve and will one day have. Too many timid souls think that freezing is complicated or difficult, whereas it is really simple and easy. Many women have said to me, "Well, yes, I'd surely *like* a freezer. But I don't know whether I can do all the things you're supposed to." That's like saying, "I'd like to dance, but I don't know if I can." All it requires is to get up on your feet while the music is playing, and start moving.

The food plan, which delivers the freezer and a few months' supply of already frozen, already packaged food into the home, gives the heretofore hesitant woman an opportunity of becoming familiar with a new procedure, thus gaining confidence. In hardly any time at all she is able to say, "Look, Ma, I'm freezing!"

For you or someone you know who may be considering the purchase of a freezer on one of the many food-plan set-ups, here is a check-list of questions to ask yourself and the salesman offering the deal:

1. Is the freezer offered a standard brand, one which you might consider purchasing even without the food tie-in?

2. Will the freezer be delivered and installed by an established local dealer or distributor who will assume responsibility for its satisfactory performance?

3. Do the manufacturer's warranties protect you against cabinet damage for at least one year and against mechanical imperfection for five years? State law permitting, is there an additional food-spoilage protection plan included in the warranties?

4. Will you have to pay no more for the freezer than you would if you bought it independently from another dealer who does not offer the food as well?

5. Does the food offered on the plan include a well-rounded selection of *the type and quality* of meat, poultry, fish, vegetables, fruit, fruit juices, prepared specialties and ice cream that you would normally buy for your family's meals and menus?

6. Who is the food supplier? Is it a local food market or frozen food locker plant which will make good if quality and quantity are not delivered exactly as stated by the salesman?

7. If there is a charge, is the food offered at definite, provable savings compared with current prices in the markets or chains where you ordinarily do your shopping?

8. Should you wish to do so, can you continue to purchase packaged frozen foods and processed meats from the same source which supplied the original order, and at similar or identical savings?

9. Are the down payment and the monthly payments no more than you would have to pay if you bought the same freezer from a different dealer and the food from other sources?

10. Is the financing arrangement made through a reputable bank or loan company known to you, and are the financing charges no greater than those permitted by law?

If you are satisfied with the answers to all of these questions, I can see no reason why you should not enter into a food-plan contract which installs a good freezer in your home to use and enjoy with profit for the whole family. Even though you may have been one of the timid novices described earlier, you will soon find yourself using your freezer with confidence and imagination, discovering new adventures daily.

Let me repeat a final word of caution, however: Do be absolutely certain that the freezer offered is fully guaranteed and particularly that its price has not been inflated to conceal the cost of the food. Printed guarantees are self-evident; you can check the price by shopping around, by looking up advertisements or by writing directly to the manufacturer.

Refrigerator-Freezer Combinations

In the formative years of the home freezing era, the so-called "freezer" combined with a refrigerator was a small, colder compartment perched across the top, not always walled off from the refrigerator itself and not always boasting a separate inner or outer door. Such a compartment did, indeed, freeze water in ice cube trays at varying speeds and might, also, have enough space left over to hold a few packages of frozen peas.

It must be remembered, however, that water freezes solid at 32 degrees. While this temperature is sufficient to provide ice cubes and even to hold solidly frozen food packages safely for brief periods, it is not cold enough to fast-freeze fresh food or hold frozen products for any length of time.

As the home freezing of fresh foods and the storage of commercially frozen products grew routinely familiar to family life, the need for more and more freezer space in compact refrigerator combination models was felt—and expressed—by the nation's consumers.

City families, especially, felt the pinch of freezer space. Their apartment kitchens could seldom be expanded to accept separate floor-standing freezers. Yet greater freezer capacity was possibly more important to city apartment dwellers than to their contemporaries who lived in private houses in suburb or country with all the space in the world for separate freezers of generous capacity. City life is busy life, and the city housewife is often also a career woman with an office job's demands added to her homemaking chores. Food shopping can be a trying and time-consuming experience in city neighborhood stores, and one which the working housewife is anxious to avoid repeating any oftener than she has to. Daily family meals must, however, be prepared and served. The solution to many city housewives' food problems—and to many others who do not live in cities but have limited kitchen space—was cheerfully and competently provided by the nation's appliance manufacturers. The freezer compartments of refrigerator combinations have been progressively enlarged until, in some models, there now exist side by side or top by bottom almost as much freezer space as refrigerator capacity.

The best of the double-duty appliances are those in which an entirely separate refrigerant system is built into the freezer compartment, one which maintains temperatures below ten degrees Fahrenheit, and insulated with denser, more efficient material than the refrigerator section. Many such models exist in a wide variety of styles.

Some combinations have a complete, large-capacity freezer joined at one side to a complete, large-capacity refrigerator. Each has a separate door, handles placed mid-center so that the two sections open like French doors. Others have the freezer compartment at the bottom instead of the top of its companion refrigerator. This makes good sense, for the refrigerator is normally used far more than the freezer during the day. Putting the freezer below the refrigerator may or may not be industrial sympathy for the housewife's predisposition to low backache, but it does eliminate a lot of bending.

Even the tiniest of kitchens have been served by manufacturers who create efficient, if proportionately small, freezer units atop their combination models. Of these, a separate outer freezer door is preferable to a freezer section whose inner door panel, usually made of thin metal or plastic, is exposed every time the refrigerator's outer door is opened.

The owner of a combination model whose freezer section maintains temperature below ten degrees can, within certain limitations, follow the directions given throughout this book for the processing, freezing and storage of fresh foods. The limitations are dictated by capacity. Under no circumstances, for example, should you ever attempt to load the compartment to capacity with unfrozen food. It is best not to freeze at one time more than three pounds of fresh or leftover food per cubic foot of freezer space. Still-frozen packaged foods from the market, however, can be stored to capacity. As a matter of physical fact, when the compartment is filled with frozen foods it will maintain safer low temperatures than it will if it is only partially filled.

Even if you use the freezer compartment of your combination model for only short-time storage, it is wise to package all foods with extreme care. Be sure that all containers, bags or wrappings are moisture and vapor proof. If a frozen store package does not have an outer wrap, slip it into a plastic bag before storing it in the freezer compartment. See Chapter 5 for a complete discussion on the importance of proper packaging.

A Few Words About Walk-in Freezers

The walk-in freezer is to the domestic double-duty refrigerator what the S.S. *United States* is to a 40-foot yacht. It's the same general idea, but bigger.

Unlike the kitchen model, the walk-in does not feature a freezing compartment on top or bottom of the refrigerator. The two units are combined side by side, or back to back, or at right angles to each other. In a few models, one walks into the refrigerator and there is another door inside which opens on to the freezer.

Sizes of walk-ins range up to refrigerators which have the dimensions of a fair-sized living room, with a freezer section of 50-cubic-foot capacity, or more. The refrigerator (or "cold room") is maintained at a temperature of from 32° to 40°F., while the freezer is capable of pulling the temperature down to – 20°F. or even lower.

A walk-in freezer is acquired, as a rule, by a farm or ranch family or by a commercial or institutional user, all of whom probably know a great deal more about freezer duties than

I do. It would be presumptuous of me to attempt instruction here. Actually, I have only one thing to say to a prospective purchaser of a walk-in freezer: Be sure the door has a release handle on the inside.

3. Your Freezer in Particular

Well, now, you have a freezer in the family. It is shaped like an old-fashioned trunk, or a Victorian highboy, or a rain barrel. It was manufactured by General Lightning, Frosby, Roger Wilco or Interplanetary Enterprises, Inc. You bought it on a food-plan deal, or paid cash for it out of your Christmas Club savings, or won it on a quiz show, or picked it up for a song from a distraught bachelor whose income tax statements fascinated the revenue agents. I don't care what it looks like, who made it or when and how you got it. You have it, and that's what I care about, because I am writing this book for you.

Your freezer is truly a family affair, for its advantages and benefits will reach out to gladden the lives of every member. Just as an automobile enriches life with the broadening joys of travel and new experiences, and just as radio and television sets bring information and entertainment into the family circle, so will your freezer make its important contributions to health, convenience, economy, relaxation and leisure.

As with most purchases in modern America, the first overtures to freezer ownership very likely began with Mother, who saw in the appliance an end to various drudgeries. Daily shopping trips, to mention one example out of many, will now become a thing of the gladly forgotten past.

Father, who pays the bills, probably began to be interested in the freezer story when he was told that food bills would most certainly decrease with freezer ownership. He may have required a little convincing, but he is a reasonable man. When that smart young salesman proved to him that savings on food purchased during seasons when they are plentiful, and therefore cheaper, would, within a comparatively short time, pay for the freezer itself and continue to issue dividends of real economy, he was genuinely interested.

Wait and see—Pop will amaze you with his participation on the family food front. For whereas the average man would sooner admit to a life of crime than to have it whispered around among his business associates and bowling companions that he helps his wife during the canning season, he will cheerfully engage in almost any aspect of the freezing procedure. Men who had to be reminded repeatedly to bring home a quarter of a pound of butter are likely to turn up in the evening with a surprise crate of freshly killed chickens or a bushel of live lobsters. The rugged gentleman who groans when a wife asks him to put the bread back in the breadbox has been known to rub his hands together briskly and inquire in a bright, interested voice, "Shall we fix the peaches for freezing tonight?" In many households, the breadwinner finds the freezer almost as provocative as Junior's electric trains.

The children will find their own pleasures and prides in the family freezer. A small son soon learns where the popsicles are kept, as well as the cookie reserves. He and a small daughter may even take turns making them. A teen-age sister entertains her fellow fan-club members graciously and effortlessly, but with full realization of the impression she is making, for she prepares dreamy dozens of little tidbits beforehand and stores them in the freezer until just before the party. While the oldest son, home from college, may disdain to take any part in food-buying, packaging or preparations, his eyes and voice will inevitably express his loving appreciation to Mother for the varied and delicious menus made possible by the freezer. And that, as every mother knows, is a pleasure beyond price or computation.

Now that you have a freezer, let's not leave it standing in the driveway in its crate. . . .

Where to Place Your Freezer

Whenever possible, your freezer should be located where it will be most convenient for you to use it. This, of course, ideally means the kitchen or very close by. Try to avoid putting it too near to the stove or in the sunniest part of a room, for surrounding heat may affect operating economy.

Be sure the floor is level and sufficiently sturdy to support the stress of a heavy weight. (A 12-cubic-foot freezer will weigh between 300 and 400 pounds empty. Filled to capacity with food it may weigh as much as 750 pounds.)

If the kitchen will not accommodate your freezer, an enclosed back porch, attached garage or a cool utility room is a good location. Lacking any of these, a downstairs "spare room" is preferable to the cellar or an outbuilding.

If you must put your freezer in the cellar, however, it is

well to observe one or two precautions. Be sure that cellar stairs are well lighted, and that there is an automatic switch which controls the *lights* from the upper landing. (The floor outlet into which you plug the freezer must, of course, work independently of the upstairs switch, or you'll be cutting power off from the freezer every time you darken your cellar.) Both precautions are calculated to safeguard against accidents, but the second one will safeguard against temper loss as well. Unless you are an accomplished juggler or have a ready assistant at hand, it is a feat to retain your balance and your good nature while you struggle with an overhead light when your arms are loaded with slippery, ice-cold packages.

Should the cellar be damp or subject to seepage it is a good idea to set the freezer on a low, solidly constructed level wooden or cement platform.

Wherever you place the freezer, however, allow a few inches' clearance between the back or side of the cabinet and the wall if yours is the type which has its condenser located in the rear or at one side of the cabinet. You can usually tell where the condenser is located by observing the exterior design of the cabinet. The spot concealing the mechanism may have a metal or mesh grill, or a louvred section, incorporated into the cabinet design.

How to Take Care of Your Freezer

A minimum of care is required, for a freezer is a marvelously engineered piece of equipment and modern finishes are hardy as well as beautiful.

For cleaning and polishing outer surfaces use any of the mild preparations you ordinarily use for your porcelain stove or refrigerator. Avoid ammonia cleaners, strong alkalies or abrasives. From time to time, wipe down the porcelain surfaces with a weak solution of your favorite household and laundry bleach to remove any stains and renew the cabinet's pristine whiteness. Rinse the bleaching solution off with clear water and polish the cabinet dry.

If your freezer has a finned-type condenser whose grilling is visible, check it two or three times a year for accumulated dust or lint. Remove this accumulation gently with a brush or whisk broom, or with the radiator-cleaning attachment of your vacuum cleaner.

Manual Defrosting (*Some models are frost-free*)

Included in some newer freezer models is a feature which automatically melts and removes frost, usually twice a day, thus preventing ice accumulation and eliminating the necessity for manual defrosting. The melting process is accomplished so rapidly that frozen food packages are unaffected. If your freezer, like mine, does not have this innovation, the amount of frost which gathers will depend on several things: the airtightness of the lid or door closure, the number of times you open the freezer, and the relative humidity of the surrounding atmosphere.

Normally, it will not be necessary to defrost completely more than once or twice a year. The inevitable formation of frost within a freezer will not appreciably reduce storage efficiency, but it may affect the cabinet's ability to freeze foods at maximum speed. An over-abundance of frost might, of course, become an obstacle by taking up valuable space which could be used more profitably for food packages.

You can keep freezing surfaces such as upright shelves, separate fast-freezing compartments and side-wall areas as clear as possible of frost by shaving it off when it is light and fluffy with a blunt plastic or wooden scraper. Do not use a sharp metal scraper which may mar the surface, and do not chip away with a sharp knife or instrument at freezer coils located underneath upright shelves. For this periodic removal of light frost it is not necessary to turn the power off or remove the stored food.

Once or twice a year, whenever you feel it is necessary and also feel ambitious, the accumulation of ice or hard-packed frost should be removed by melting. The best possible time to defrost would be during a dry, cold winter just before you plan to restock the freezer with a supply of fresh meats, or on a cool autumn day before loading the freezer with the fall harvest of fruits and vegetables.

In other words, try to plan your all-out defrosting to coincide with the time of the year when your frozen food supply is at its lowest.

Procedure for Complete Defrosting

1. Remove all frozen food packages from the freezer. Put as many of the meat packages as possible in your refrigerator, which should be adjusted to its coldest temperature. Place overflow packages in heavy cardboard or wooden boxes. Line the cartons or boxes with several thicknesses of newspaper and cover the food packages with a blanket or with additional thicknesses of newspaper. The tightly packed, well-covered frozen packages will keep each other cold during the comparatively short time it takes to complete the defrosting job. If you are defrosting on a cold winter day, you may store the cartons temporarily out of doors in a protected place.

2. Turn the Temperature Control to the "off" position, if this position is indicated on the knob; some manufacturers omit it to prevent accidental turning off of the motor. If there is no "off" position, turn the knob to its lowest register. In either case, detach the cord plug from the electric outlet.

3. With the blunt paddle described previously, scrape as much frost as possible from the surfaces. Do not work too

hard at this, as the only purpose is to reduce the amount of ice which must be melted.

4. (a) You may leave a chest lid or an upright door open and wait for the warmer air of the room to melt the ice—or—

(b) A preferred, faster method is to place pans of fairly hot (not boiling) water in the bottom of a chest or on the shelves of an upright, and close the lid or door. Within a very short time the ice and frost will release their hold on sides, bottom or shelves and can be easily coaxed out in large pieces.

It is not necessary, or even desirable, to wait around until the ice has melted into water. Put a rug of thick newspapers around the outside base of the freezer to catch drippings and keep things neat. If your freezer is located near a sink or laundry tub, remove the pieces of ice as they become detachable and toss them into the sink, where they will melt. If not, put them in a large bucket or container placed handily near the freezer.

5. When all ice and frost are removed, some condensed moisture will remain. Use absorbent rags or, better, a large clean sponge to draw up this excess moisture.

6. While the freezer is empty and newly defrosted is a good time to do a thorough interior cleaning job with a mild solution of warm water and baking soda (3 tablespoonfuls soda per quart of water) or with clear warm water to which you have added a tablespoonful of your household deodorant, provided this is a gentle one which *does not leave a strong smell.* Do not use ammonia or caustic alkalies. Rinse with clear water, then dry the interior of the cabinet and all shelves thoroughly with a lintless towel or cloth.

7. Plug the cord back into its electric outlet and turn the freezer temperature control knob to "on" in its coldest position. Close the lid or door and allow the empty unit to run

for at least half an hour to bring down the temperature inside the cabinet.

8. Return frozen food packages to the freezer, wiping them off with a soft cloth if moisture has formed on them. Take inventory of your supply at this time, especially if you maintain a dated log or file of the foods you have frozen in the past. Put newer packages at the bottom or back of the freezer, keeping the older ones more readily available for early use.

What to Do in the Event of Non-operation

In an excellent leaflet prepared in consultation with the United States Bureau of Human Nutrition and Home Economics (#321, U.S. Department of Agriculture Extension Service, Washington, D. C.), Miss Evelyn L. Blanchard very wisely advises home freezer owners to be prepared for emergencies. She suggests, in part, that you:

1. Find out about your nearest locker plant.
2. Try to locate a good source of dry ice in your community.
3. Keep canning supplies on hand and keep canning equipment in good working order.
4. During the seasons when power failure may be frequent in your community, run your freezer at its lowest temperature.

These are practical suggestions, especially the last one. Your freezer is so well constructed and insulated, however, that barring an act of God or war, operating failure is not likely to extend to the point where your frozen food supply is in danger. Most freezers have such foolproof mechanisms that breakdown is extremely rare. The following precautions and instructions are included just in case.

If your freezer fails to operate at any time, *don't worry.* Check first for simple causes. (I once knew a young man who fancied himself as a first-class mechanic, claiming a keen knowledge of electricity among his talents. One evening, a favorite lamp in my living room failed to function and before I could say "James Watt" he had the lamp apart and the floor plug nakedly exposed and was calling for friction tape in the urgent tone of an operating-room surgeon saying "scalpel." Quite a while later, after the lamp was put back together and the floor plug reassembled, he reached over with a triumphant gesture to switch the lamp on. Nothing happened. Mumbling an apology, I unscrewed the bulb and replaced it with a new one. The lamp lit.)

If therefore, your freezer is not freezing and you have glanced at the control switch to make sure it was not accidentally moved to "off," examine the cord plug and ascertain if it is properly in the floor outlet. Something may have jarred the plug loose and the prongs may not be making proper contact.

If the plug is tightly in the outlet but the freezer still does not work, remove the freezer plug and test the outlet with a small lamp or fan.

If these do not work, have someone take a look in the fuse box.

If plug, outlet and fuses are apparently in order, check the lamps and appliances in the rest of the house to discover whether there is a general power failure in your neighborhood.

If this proves to be the case, try to find out from your local electric company approximately how long it will take to repair the power lines. If you are assured that it will be only a matter of minutes, or a few hours, or even a couple of days, it may not be necessary to do anything except to keep the freezer lid or door tightly shut to maintain interior cold. I say it "may not," because this depends on the size and insulation of your freezer and the amount of food it contains.

With door or lid closed, a well-constructed, well-insulated six-cubic-foot freezer will keep a capacity load of food safely frozen for as long as two days; a 12-cubic-foot freezer for three; an 18- to 21-cubic-foot freezer for four or five days. The more food there is in the freezer, the longer it will stay frozen. A freezer that is half filled will stay frozen about half of the periods of time mentioned above.

On the other hand, if the electric company is vague about the length of time it will take to make repairs, and you feel a little nervous about all that food in the freezer, perhaps it would be wise to play safe and arrange for the delivery of a 25- or 50-pound piece of *dry ice.* If you have been beforehand with your preparedness for this eventuality, you probably know the locker plant, ice company, ice creamery or dairy in your community which has dry ice for emergencies.

Wearing heavy gloves (dry ice can cause skin burns) chop the ice into comparatively small pieces and distribute them evenly over the package foods inside your freezer, first covering the packages with layers of cardboard, heavy thicknesses of newspaper, thin wooden boards or sheets of metal.

After twenty-four hours, if the power is still off, repeat the dry ice treatment as before.

It may be more convenient for you to avail yourself of the facilities of a community locker plant, unless the manager there is in the same fix that you are—powerless.

Should you find, however, that there is no general power failure in the vicinity and no fuse or conduit trouble in your house, you may accurately deduce that the fault lies in the mechanism itself.

Call your dealer or service man immediately and meet him at the door with the manufacturer's warranties clutched in your hand.

Unless you are satisfied that the difficulty can be repaired at once, transfer your food to a locker plant; if there is no locker plant near you, perhaps you have some good neighbors with some space to spare in their freezers. If neither of these courses is open to you, impale the dealer or service man on the determined glint in your eye and inform him in no uncertain terms that he can jolly well see to it that a substitute freezer be rushed to your place right away to house your endangered food until your own cabinet can be put in working order again.

In the event that you discover the non-operation of your freezer after some time has elapsed and you suspect that some of the food has thawed either wholly or partially, start making plans for a day of wholesale cooking and canning.

It is not wise to refreeze thawed raw foods, except fruits whose flavor has not been spoiled by fermentation. Even these it is better to cook or use for jams and jellies, because their flavor suffers if allowed to remain long in the thawed state before refreezing.

Thawed food *which has not spoiled* can be refrozen after it has been cooked, however. Meats can be braised, roasted, broiled, panfried or made into stews and frozen when they have cooled. Vegetables can be cooked until almost done and stored for quick meals.

Do not under any circumstances refreeze fish and shellfish which have thawed completely, and never eat frozen cooked leftovers which have been thawed for any length of time. The exception to this is baked products, which may be eaten even though completely thawed so long as they taste good.

If you live in a region where power failure is a frequent occurrence, you should inquire about a food insurance policy at your local locker plant, bank or insurance company.

4. First Steps Toward Freezing

Your freezer stands in your home as a symbol of convenience, economy and more gracious living. As time goes on, you will come to realize more and more its benefits to you as a homemaker and to the entire family; you will, increasingly, lean heavily on its amiable service and its consummate resources; you will find yourself wondering how you ever managed to get along before it came into your life. Eventually, as with all phenomena grown familiar, you will take it as much for granted as you do the miracles of electricity and air travel.

Your freezer's tangible contributions will be fully recognized and appreciated as you learn to use it with maximum efficiency. There will always be cookies and treats on hand for the children; drop-in guests can always be served an impromptu, delicious and effortless meal; you will be able to take a week or two off, if you feel like it, secure in the knowledge that your freezer is hoarding dozens of complete, well-balanced meals.

There is, however, one very important service performed by your freezer which does not show itself immediately or in any way that you can see it, yet this is the most dramatic role it can play in your life. I call it the "soul of the freezer," and I hope to be forgiven for any attribution of emotion or sentiment to a mechanical object.

The soul of the freezer is its ability to provide for you and your family food of so much better quality and nutritive value that it can, if you permit it, make life a more joyful experience through increased energy, freedom from illness and a buoyant feeling of positive well-being.

That is a large expectation to anticipate from a manufactured cabinet composed largely of cold metal and enamel paint and concealing a motor-driven mechanism of modest proportions. It is, however, an expectation which can be fully realized.

It is entirely true that you will never take from your freezer food of better quality than you put into it. It is equally true that another marvelous mechanism, your body, will, within the limits dictated by your heredity, serve you in direct proportion to the life-giving qualities of the food you put into *it*.

Much has been said and written about the importance of balanced diet. We know, or think we do, all about vitamins. Food research engages the full time of hundreds of thousands of government, education and business personnel, and results of their research reach us in the form of pure food laws, new food varieties and enriched food products.

In the scientific journals of the medical profession there is mounting evidence that our doctors and dentists are paying increasingly more attention to the prevention, treatment and cure of disease through diet therapy. The very term "nutrition" is being written into the language as the title of an emergent science.

It is generally believed that the people of this country are better fed than are people in any other portion of the world. We think of ourselves as a healthy, robust nation. Yet, as the serious nutritionists point out, the country does not have nearly enough hospital bed space to accommodate the number of people who need hospital care. They remind us that the nation buys more than a hundred million dollars' worth of laxatives every year, and that approximately 40 per cent of drug store income is derived from the sale of stomach remedies of one sort or another.

The statistics on the incidence of illness and disease in our nation are alarming, and need not be quoted here. The fact remains that too many people are troubled by too many discomforts to which they have resigned themselves, thinking that such discomforts are synonymous with being human. There are a number of good books written by serious nutritionists who describe at length the pitfalls of modern American diet and suggest ways to avoid or overcome the deficiencies to which this diet leads.

The word "nutritionist" is vastly overworked, and I wish to point out that I do not mean "food faddist." The honest worker in nutrition does not claim miracles or attribute to any one food or group of foods the ability to cure disease.

Diet deficiencies, we are told, result in large part from the inherent weaknesses of the very food we eat.

Our bread is baked from flour which has been refined and processed to make the product aesthetically appealing, and we have acquired a universal taste for it. The refining process, however, removes from wheat its natural nutrients and replaces them with nice-looking, nice-tasting bulk which has lost much of its food value.

Rice is bleached and polished until it is a joy to behold, easy to cook and pleasant to eat. A few of the essential food elements are then put back into the white bread and (in some cases) white rice, but the greatest portion of them is carefully collected and sold to pharmaceutical companies who manufacture the vitamin pills we later buy because the vitamins were removed from the bread and rice in the first place.

Frozen foods—those you buy and those you process yourself—can do a great deal to raise your family's diet level to a more healthful standard.

Frozen Foods Compared with Fresh or Canned Varieties

The fresh fruits and vegetables you used to serve your family before you acquired your freezer were often dependent upon the season of the year. You reveled in spring strawberries, late summer corn, autumn pumpkins and winter squash. Fresh leafy green vegetables, however, appeared on your table with praiseworthy regularity no matter what season, for you knew that these were foods of vital nutritional value.

The marvels of modern transportation made it possible for you to buy spinach, broccoli or asparagus all year long, for our national climate varies to such an extent that it is always harvest time somewhere in the United States. But the "fresh" spinach you bought in a season when your own regional farmlands were crusty with frost may have been a week to twelve days old by the time it was carried home in a shopping bag and set to cooking on your stove. By that time, it had probably lost up to 65 per cent of its vitamin content.

Vitamins are volatile and temperamental. They thrive on air and sunlight while the food is still attached to a living plant; they start to escape on exposure to sun and air as soon as the food is plucked from tree or vine. Then, unfortunately, we are apt to see the very last of the more tenacious vitamins gurgling down sink drains in the cooking water, for the reprehensible crime of over-cooking vegetables in too much water is committed every day in American kitchens.

In canned food varieties, a good portion of the original nutrients is preserved, for the great canning companies usually locate their plants close to the crop sources, protect the foods from light and air, and control cooking temperatures with scientific measures. As we know, however, the taste and texture of the original food are altered during the canning process.

The commercial freezing companies, too, establish their processing plants close to crop sources and, like the canners, make a sincere effort to select locations where the soils are fertile and rich in minerals. Many of the large food-packing firms conduct exacting soil tests and replenish the earth with minerals when needed to improve the harvest. They maintain strict surveillance over the quality of the seeds or shoots planted for crops. They are responsible for developing many new, improved varieties of more savory taste and greater nutritive quality.

In using your freezer for the freezing and storage of fresh fruits, vegetables, meat, poultry, fish and dairy products, it is wise to take a page out of the commercial companies' book and remember that every fresh food item has its peak season, when quality and flavor are at their best.

The small amount of extra care you take in choosing foods when they are at their nutritional summit will pay off at a tremendous rate of interest, for these foods will taste better and be better for you.

Resist the temptation to buy a few bushels of late-season fruits or vegetables offered by local markets or farms at special prices. Learn, instead, the times when the different varieties are at their prime of young maturity and do your bulk purchasing then. The same rule applies to meat and poultry products, and to dairy foods.

An attempt will be made throughout the pages of this book, in the sections pertaining to each food item, to reflect the newest information available from government and educational experimental stations as to the varieties most suitable for freezing, the seasons when they are at their prime in greatest quantity, and the recommended methods for preparing them for the freezer.

The Importance of Quick-Freezing

Your own observation has taught you that some foods "spoil" more quickly than others. The spoilage is evidenced in various ways and caused by more than one factor, but the result is always the same—the food becomes inedible. A glass of milk left uncovered in a warm room will turn sour within a short time. A peach under similar conditions will become discolored and brown. A piece of bread, however, will take considerably longer to show evidence of its deterioration.

All fresh food is alive, whether it is still on the hoof or in the garden, or whether it is in your refrigerator or vegetable bin waiting to be cooked. That cooking itself does not "kill" food is evidenced by the fact that even after you eat it the vital elements it retains nourish your body's cells. The philosopher who stated that "you are what you eat" spoke an unarguable truth.

Although some foods deteriorate more rapidly than others, and consequently the rate of speed at which they must be frozen varies with their individual temperaments, a good rule for home freezing is—the faster the better!

When perishable food is not preserved by any one of the known means such as canning, drying, pickling, smoking and of course freezing, enzymes within the cells of the food continue to live out their respective spans at their normal pace. As they do this they effect chemical changes which, after a certain point has been reached, become destructive to the physical properties of the food. Fruits become squashy and unpleasantly fermented, vegetables become progressively limp, pale and shrunken.

Enzyme is the name given to a classification of proteins about which not very much is known. That is, it is not known exactly what they are; it is known what they do, however. They act as catalysts which accelerate chemical reactions. Enzymic action is a life progression. It is present from the germination of a seed throughout the period of maturation to the moment—and beyond it—when you throw a rotten old apple or Brussels sprout into the garbage container.

In animal products, the degeneration is even more dramatic, for along with enzymic change there is also enacted a form of microscopic bacteriological warfare. Invisible little creatures (their biological name is "microorganisms") feed on the animal-derived food and eventually render it unfit for human consumption.

The preservation of food at the low temperatures made possible by freezing slows down enzymic action and *completely checks* the growth and reproduction of destructive bacteria. The faster food is frozen, therefore, the sooner both of these undesirable processes are rendered harmless.

But something else happens during the freezing process, and it is this something else that, added to the quality of the food you put into your freezer and the way you package it, makes the end result either very good, just passable or downright disappointing.

Cold, we were taught in chemistry classes if we were paying attention at the time, is merely the absence of heat. In our courtrooms, the accused is innocent until he is proved guilty. In nature, physics, chemistry and refrigeration, everything is hot until it is made cold by a process of heat extraction.

Of course, both "hot" and "cold" are purely relative terms. It may be cold out today, but it was colder yesterday. Tomorrow, we hope, it will be warmer. Next August, presumably, we will have the hottest day of the year. The weather reports will so record the fact, which will make newspaper headlines. They will mean only that the day was hotter than any other one of the preceding 364. If the day is reported as a record-breaker, however, it will mean that this particular day was hotter than any day of similar date in recorded history.

Freezing, too, has its relative degrees. Because just about every kind of food you or I will freeze has as one of its component parts a measurable amount of moisture, or water, when we speak of freezing food we speak of it in terms of the freezing point of water.

Water, as we know, freezes at 32° Fahrenheit. It would seem reasonable to assume, therefore, that when the moisture content of food reaches a temperature of 32°, it will be frozen. By "freezing," we mean that water has changed from a liquid state into a solid.

To our surprise when we first observed the phenomenon, we observed that apparently a given amount of liquid freezes into a *bigger* amount of solid. This is old hat to even the youngest chemistry or physics student, but I remember that I discovered the principle one cold winter morning when, as a little girl, I reached for the bottle of milk on the doorstep. The milk was frozen, and some of it thrust up from the top of the bottle, displacing the cap. This was a rewarding discovery, because the topmost part of the frozen milk tasted very much like ice cream.

If I had been more interested in scientific research at the time, I could have conducted a simple day-to-day test. After a while, if the weather cooperated by varying its temperatures from 32° to 0°, I would have found that the milk stalagmite was tallest on the days that hovered around 32°, shortest on days that got down below zero. Yet, always, it was the same amount of milk—one quart. How come?

To one degree or another, all perishable food contains moisture. Some of this moisture exists as water, some as a solution of water and something else. My bottle of milk was composed largely of water—87 per cent. The rest of its composition—all that good calcium and the important vitamins and milk proteins (including enzymes)—existed in the form of a fairly consistent solution, or emulsion.

Because milk is a liquid we can actually see how it solidifies as it freezes. What happens is that as the temperature of the surrounding air is lowered beyond the freezing point of water, the water progressively crystallizes out in the form of pure ice. This process leaves behind it increasingly concentrated solutions of the other substances in the emulsion, solidifying these at *their* freezing points. The size of the ice crystals which form is controlled by the span of time during which freezing takes place. If the temperature is lowered slowly, the crystals expand considerably. If freezing is sharp and sudden, the crystals retain approximately the same size as the original moisture molecules.

When foods like soup or juices are frozen from a liquid form to a solid, the expansion of the crystals is not too important. Once the liquid is thawed or heated, the content melts back to its original volume and may be stirred or shaken back into emulsion.

In the more "solid" foods, however, such as meat, fish, poultry, vegetables and fruits, the size of the crystals which form during freezing plays a vitally important part in determining the quality of the food after it is thawed for use.

As meat freezes, for example, the moisture within its tissues forms into ice crystals. If these crystals expand to a large degree by virtue of slow freezing, they occupy more space. In so doing, they puncture and destroy the surrounding walls of the tissue cells. This "breaking down" of tissues permits the remaining moistures, or meat juices, to migrate from their original molecular suspension. The ultimate result is a decided loss of flavor.

This insidious process of cell destruction through slow freezing is even more dramatically evidenced in fish, much of whose moisture is contained in the form of unsaturated fish oil. If fish is frozen slowly, a condition results which we recognize as "rancidity." (See page 93.) This happens because the unsaturated oils, which freeze at a much lower degree and more slowly than water, have leaked out into the body of the fish, permeating the flesh with an unpleasantly oily taste.

Slow-frozen fruits and vegetables, depending on their cell structure, are likely to become limp and flabby as soon as they are thawed.

Baked goods, on the other hand, because they do not have much moisture content do not depend so much upon fast-freezing in order to retain their equivalent fresh texture and quality.

For the purposes of this book, which is to direct you to rewarding results via simple procedures, it is not necessary to go into lengthy details about the various temperatures at which the various foods reach their solid-frozen points. It is important, however, to know how to apportion the freezing space in your freezer to get invariably successful results. Further, it is extremely important to know *how much* food to freeze at one time.

Both of these considerations depend upon the style and

brand of your freezer, and whether or not it has a separate fast-freezing compartment if it is a wrap-around chest or upright. Each type will be discussed separately, and you can follow the recommendations suggested for the style of freezer you own.

WRAP-AROUND SIDE-WALL CHEST (without separate fast-freezing compartment):

1. Set thermostat control to its coldest position at least two hours before placing packages in the freezer.

2. Place packaged food in contact with the side walls of the freezer, allowing a little space between the packages for air circulation. If your model is described in the manufacturer's literature as having heavy freezer coils on *five sides* of the cabinet, packages will also freeze quickly when placed on the floor of the freezer.

3. Do not stack warm (unfrozen) packages on other unfrozen or frozen packages.

4. Multiply the cubic capacity of your freezer by four pounds. This is the maximum amount of food to freeze at one time. For example, if you have a ten-cubic-foot chest, it is better not to freeze more than 40 pounds of newly packaged food at one time. If you have more than this amount to put in the freezer, hold the remainder under refrigeration until the first batch is hard-frozen (allow at least 18 to 24 hours for this). This applies to meat or poultry. It is best not to buy any more fish, fruit or vegetables than your freezer can handle in one day.

5. After the packages are thoroughly frozen, they may be moved to the center of the freezer to clear the side walls for the next batch of food.

SIDE-WALL CHEST (with separate fast-freezing compartment):

1. Set thermostat control to its coldest position at least two hours before placing packages in the freezer.

2. Place packages in the "sharp-freeze" section or on the shelf designated by the manufacturer for this purpose.

3. Spread packages on the floor and against the side walls of the compartment, allowing a little space between the packages for air circulation.

4. Do not stack the packages on each other, but spread them in a single layer.

5. Place as many packages in the compartment as it will hold, unstacked, at one time.

6. Do not waste "fast-freeze" compartment space by filling it with packaged cooked leftovers or with baked goods. These can be placed immediately within the chest, the leftovers against the side walls and the baked goods in the center of the cabinet or in the baskets provided by the manufacturer.

7. After the food is thoroughly frozen, it may be transferred to the chest itself.

WRAP-AROUND UPRIGHT (without freezer plates):

1. Set thermostat control to its coldest position at least two hours before placing packages in the freezer.

2. Place packages on shelves but *against the back and side walls* of the cabinet, allowing space between the packages for air circulation.

3. Do not stack the packages, but arrange them in a single layer.

4. Multiply the cubic capacity of your freezer by four pounds. A 15-cubic-foot wrap-around upright will safely hard-freeze up to 60 pounds of fresh food at one time. Hold additional amounts of meat or poultry you want to freeze under refrigeration until the side-wall space is cleared.

5. After the food is thoroughly frozen (allow 12 to 20 hours) the packages may be stacked on the shelves toward the center of the cabinet.

UPRIGHT (with freezer plate shelves):

1. It is not necessary to reset the thermostat control, provided this is already in its normal position or unless a capacity amount (see 4, below) is to be frozen.

2. Place packages on freezer plate shelves, allowing a little space between the packages for air circulation.

3. Arrange packages in one layer, without stacking them on each other.

4. Multiply the cubic capacity of your freezer by seven pounds. A well-insulated 12-cubic-foot upright freezer with freezer plate shelves will fast-freeze up to 85 pounds of food within 12 to 14 hours; an 18-cubic-foot upright of similar construction can handle up to 125 pounds in the same length of time.

5. After the food is completely frozen, stack and "file" it on the shelves in any way that is convenient to you.

6. Whenever possible, leave one shelf or part of it clear of stacked packages so that fast-freezing space will always be available, thus eliminating the necessity for undue handling of frozen packages.

7. Unless the bottom shelf of your freezer is specifically designated by the manufacturer as a fast-freezing shelf, it is best to use this space for the storage of already frozen foods, cooked leftovers and baked goods.

ALL STYLES

After the freezing job is done and the packages rearranged, the thermostat control may be set back to "zero," "normal" or "storage," whichever it is called by the manufacturer of your freezer; this designates the temperature at which frozen foods should be held for safest long-term storage.

How to Arrange Frozen Foods in Your Freezer

After foods are frozen, especially if they are wrapped or contained in opaque packages, it is easy to forget just what is what, and where it is.

Inasmuch as the most economical use of your freezer depends on your turning over the supplies it contains to make room for new seasonal purchases of the same varieties, you will want to avoid "losing" packages in the murky depths of your chest or in the far back recesses of your upright. By the time you find them again, they may have passed the peak of their savor, texture and freshness, for there are certain limits to the length of time the various foods may be safely stored. This subject will be investigated in the specific food chapters to follow, and the recommended storage period for each food item will be listed.

Inevitably, you will "freeze and forget," temporarily, some food varieties during the seasons of their greatest abundance. For example, it would be silly to rob your freezer of newly frozen strawberries in May and June, when they are cheap and plentiful. At this time of the year you will serve fresh, unfrozen berries to your family and save the fresh, frozen ones until fall and winter, when they are at a premium in local markets.

One good idea is to maintain a perpetual inventory of the foods you freeze and store, entering quantities and dates frozen on a log sheet or in a notebook and remembering to deduct the number of packages removed from the freezer as they are used for family meals.

Such a log, called an "Inventory of Frozen Food and Use Record," has been prepared by Nancy K. Masterman and Victoria M. Chappell of the School of Nutrition, Cornell University. According to the Cornell List of Publications extension bulletin, these log sheets may be purchased for a few cents by addressing the Mailing Room, New York State College of Agriculture, Ithaca, N. Y.

Some commercial printing companies also distribute freezer log sheets or books at nominal cost. You can, of course, make your own with a ruler and a sharp pencil or a fountain pen, listing food items alphabetically at the left side of the paper, with the number of packages frozen for each item stroke-tallied alongside the various classifications. If you freeze any one food more than once a year, it is wise to indicate the month it was frozen so that you can be sure to use the older packages first.

Another good idea, especially if you have a chest type freezer, is to "make a map" of the contents on a large piece of wrapping paper which you can freezer-tape to the underneath side of the lid. This can be used in conjunction with your log book, or you can list the number of packages by stroke tally directly on the "map," marking them off with a different colored pencil as they are removed from the freezer.

When putting food away in the chest freezer, separate all meats by variety and cut, i.e., *Beef:* roasts, steaks, hamburger; *Pork:* roasts, chops, sausage, etc.

As far as it is possible for you to do so, allocate permanent space in the freezer to each variety, placing heavy roasts, hams, turkeys, etc., at the bottom, smaller cuts on top of them, and easily moved containers of fruits, vegetables, prepared foods, and so forth, on top of *them.*

If you prefer, and if you can predict with fair accuracy a month of meals in advance, you can arrange foods in sectional layers in the order in which they will probably be used. This latter system will permit you to reach the foods as they are required without a lot of digging and disarranging of other packages.

Most chest freezers come equipped with non-corrosive

metal or plastic baskets which fit across the topmost part of the cabinet. (If your freezer does not have any, they can usually be purchased separately from hardware or appliance stores dealing in freezers and accessories.) It is wise to use these baskets for lightweight or soon-to-be-used foods, for two reasons. First, if baskets are filled with heavy foods, they will be hard to lift or slide out of the way when you want to reach into the lower depths of the cabinet. Second, and more important, the heavy foods are generally packaged meats, and these should invariably be kept in the coldest part of the freezer to assure their long-time quality. Located as they are near the top of the cabinet, baskets are likely to be in the area of "warmest" cold, for heat rises. In a well-made, well-operating chest, a thermometer should register 0°F. two inches below the lid contact. Your guarantee probably specifies this. If you suspect that your freezer is running warmer than it should, check it with a thermometer. If your suspicion is verified, call your dealer's service department. A simple adjustment can usually correct the fault.

An upright freezer is somewhat more convenient, for you can keep the various food items on different shelves for greater accessibility. In my upright, for example, I store beef and veal on the top shelf; pork, poultry and seafood on the second; variety meats and vegetables on the third; fruits, leftovers, soup stock, eggs and dairy food, pre-cooked recipes, bread and rolls on the bottom.

What Foods to Keep in Your Freezer

It would be presumptuous of me even to suggest a list of the foods you "should" keep in your freezer, for such a selection depends entirely upon your personal preferences and those of your family. You will note that the enumeration of the items kept in my freezer omits entirely such things as lamb, mutton, cakes and pies. The reason for the omission is that I have a personal dislike for lamb in any form, and rarely eat pastry desserts.

When I entertain guests who like sweets, I make or buy something special for the occasion. The freezer is put into useful service at these times, too, for very often the dessert is prepared or purchased at my convenience days ahead of the dinner party, wrapped in foil or placed in a polyethylene bag, and parked temporarily near a freezing surface until an hour or so before I plan to serve it.

The closest I will come to making any suggestions about your freezer inventory, therefore, is to remind you that meat is the most expensive item on your food budget, and that perhaps you should allow a good 60 to 75 per cent of all your storage space for quantity purchases of your family's favorite meats.

Besides, you will learn through experience which of your favorite frozen foods are consumed by your family most quickly and will get into the habit of calculating your seasonal purchases accordingly. In my house, for example, I will deliberately by-pass pie apples or green beans, preferring to leave as much freezer room as possible for green peppers in the fall, mushrooms and asparagus in the spring. During July and August I am likely to devote a whole shelf to sweet corn which, frozen on the cob, takes up an awful lot of space; but it is one of my favorites of any season, and therefore worth it.

What About Refreezing Thawed Foods?

Don't.

To be on the safe side, don't even refreeze *partially* thawed foods if you can help it, unless you cook them first.

I realize some pamphlets and books on the subject of home freezing have scoffed at the idea that any danger exists in refreezing foods which have defrosted. I realize also that my peremptory negative injunction eliminates some opportunities of buying frozen foods in large economy-size packs, allowing them to soften until they can be handled and repackaging them in portion-sized containers. I can't help it. There are exceptions to the rule, but unless you are pretty sure of what you are doing, refrozen foods may lead to upset stomachs or worse.

In the discussion at the beginning of this chapter, we explored the desirability of quick-freezing and learned that low temperatures retard enzymic action and check microbial growth and reproduction. This is why frozen foods remain fresh and unspoiled throughout their recommended storage periods.

When food is thawed, the retarded enzymes burst into activity again and the arrested bacterial invaders, released from imprisonment, celebrate by reproducing themselves like crazy. While it is true that the extent of bacterial multiplication depends on the food item itself and also on the time lapse between the frozen and the thawed states, the whole subject is so tricky that it is better to err on the side of too much caution than too little.

Sugared berries or cherries, for example, do not seem to house great numbers of microorganisms, and may be refrozen so long as fermentation has not made much inroad and they are still palatable. Drier varieties of baked goods such as breads, rolls and plain cakes can be thawed and refrozen as many times as you please.

When it comes to meat products, poultry, fish and most vegetables, however, it is another story; and unless you have an expert's knowledge of food chemistry it is better not to fool around.

It is, therefore, not particularly wise to buy bulk-packaged frozen meat, fish or seafood *unless you chop or saw off* smaller portions and repackage these while the food remains solidly frozen. Large institutional packs of frozen peas and lima beans are an economy if they are dry-packed so that you can pour them off into smaller containers without defrosting them. Some frozen food producers offer institutional packs which are composed of a number of meal-sized portions, individually wrapped. If you buy these, check the individual wrappings before tucking the separate packages away in your freezer. You may want to protect them with a good outer-wrap of moisture-vapor-proof material.

Perhaps you think I am being unduly alarmist about this thing. In the *American Journal of Public Health* some years back it was reported that a gram of beef steak had a count of 390 bacteria when the meat was in its frozen state; after twenty-four hours at 70°F., the count had risen to 1,400,000. Oysters, similarly tested, rose from 22,000 per gram to 320,000,000.

5. Wrap It Right . . . or You'll Regret It!

The importance of proper packaging of all the foods you place in your freezer cannot be emphasized too much. You mustn't think, however, that skillful packaging is difficult. Every phase of the freezing procedure is easy, including packaging. It is stressed here because of its known and measurable influence on the success or failure of the end product, the food you bring to the table.

The marvelous efficiency of your home freezer and the superb quality of the foods you take pains to preserve for later enjoyment can be canceled out ruthlessly by careless wrapping or by the use of inadequate materials. On the other hand, informed precaution as to the selection of materials and the meticulous practice of expert packaging can actually extend the storage life of foods within your freezer.

As with most things in this life of opposites, there is a right way and a wrong way to do the job of frozen food packaging. It is just as easy to do it right.

In order to understand why and how foods are influenced by the manner in which they are wrapped, perhaps we should take a look at what goes on in the freezer after a food package is delivered into its chill recesses for freezing and storage.

In a manner of speaking, your freezer itself is a package; it is a big box full of dry, frigid air. It was specifically and adroitly designed like this, for cold is merely the absence of heat and moisture contains heat until it reaches the solidifying, or freezing, temperature.

When a package of food is placed in your freezer, a sort of tug-of-war goes on between the heat within the package and the surrounding cold, dry air. In a way, this is a full-scale battle of the elements. Because heat itself is nature's preferred condition, the cold air attempts to draw warmth, consequently moisture, from the food. Its ability to draw warmth is in your favor, because that's what freezing is. Its ability to draw moisture, however, unless prevented forcibly from doing so by a barrier of some kind, is decidedly not in your favor. If the cold air wins this battle, the result is *dehydration* of your food.

Dehydration is sometimes called "freezer burn," and it occurs most frequently on improperly wrapped meats, fish and poultry. Perhaps you have seen a steak, roast or chicken with discolored grayish-brown areas on the surface after it has been removed from a freezer (not yours, of course!). Cold air won the skirmish here, maliciously removing essential moisture and juices from the meat.

Dehydration is objectionable not only because of the unappetizing appearance of the meat, but also because the migration of juices has rendered the meat dry, tough and several degrees less than tasty. Dehydration within a freezer is not immediate or rapid, but the tug-of-war continues day after day, month after month, long after the food has been fast-frozen and stored away.

It is essential, therefore, that meat products especially be protected by wrapping materials which resist the transmission of moisture-vapor to the surrounding air. This protection is called a "moisture-vapor barrier," and some freezer packaging materials have it while others do not. (A complete discussion of the various types of freezer wrapping materials available to you will be found toward the end of this chapter.)

Another hazard of faulty and inadequate wrappings or containers is *oxidation.*

Oxidation, erroneously confused in many people's minds with dehydration, is a different process altogether. Whereas dehydration moves *outward* from the food package—that is, moisture is robbed from the food and "breathed out" into the surrounding air—oxidation moves *inward.* In this action, oxygen from the air invades the frozen packages, mingles subversively with the food therein and ultimately commits a dastardly crime: It creates rancidity in meats and fish and steals valuable vitamins from fruits and vegetables.

What you and I generally describe loathingly as "rancidity" is actually the formation within the fat-containing cells of meat of fatty acids, peroxides and aldehydes. They, in turn, are the result of contact between good old oxygen and food fat cells, and they impart to the meat or fish a disagreeable odor and taste.

The higher the fat content of meat or fish is, the more vulnerable it is to oxidation, or rancidity. It is for this reason that most freezing authorities warn against cold-storing frozen pork and pork products for more than six months. Pork is not only heavily larded with thick layers of fat, it is also permeated with millions of tiny fat cells, far more than beef, veal, lamb or poultry. Although the following statement will be repeated and enlarged upon in our later discussions about the storing of foods, it should also be mentioned here: There is no reason why pork cannot be stored at 0°F., if desired, for periods of time far longer than six months *provided it is properly packaged and quickly frozen.*

The freezer wrappings which protect against oxidation, therefore, must be impermeable by the oxygen present in the freezer's interior air. This protection is called an "oxygen barrier," and, like the moisture-vapor barrier, some packaging materials have it while others do not.

While dehydration and oxidation are the two most serious hazards which confront food in the freezer, these bugaboos are not the result of careless wrapping alone, or of materials which lack effective oxygen-vapor-moisture barriers. Fractures or punctures in the packaging materials also exact their toll.

External breaks or cracks in food packages or containers, of course, are very likely to result in dehydration and oxidation within, for obvious reasons. Such breaks or cracks can be caused by a number of things: Rough handling, sharp edges inside the package (such as bones), brittleness of the wrapping or container, or filling a container too full to allow for expansion.

To safeguard against the occurrence of breaks, therefore, it is necessary that the wrapping material or container be very strong, thus reducing the possibility of puncture. In addition to strength, sheet wrappings must, or certainly should, be pliable so that they can be molded closely to the mass of food, excluding air pockets (which contain moisture and oxygen). In technical language, this moldability of paper is called "deadness."

Internal fractures of coated sheet wrappings, less easily seen than the external breaks or cracks, also represent a threat to the packaged food. Such coatings are applied during manufacture by various methods to many of the materials available on the market. They form a valuable protection so long as they are not pierced during handling or by brittleness at extremely low temperatures.

To sum up—when shopping for freezer packaging materials which insure protection against dehydration, oxidation and external or internal punctures, look for meat wraps plainly labeled moisture-vapor-proof, oxygen-resistant, flexible, and strong. Containers for fruits, vegetables, dairy products and cooked foods should also be moisture-vapor-proof and strong, and should possess the additional qualities of being easy to fill, and shaped for economy of space within the freezer.

Any packaging material used in the freezer must, of course, be sanitary, rust-proof, grease-proof, odorless and tasteless.

Packaging Materials for Meat, Poultry, Fish

No attempt will be made here to discuss the quality or desirability of the many meat varieties and cuts, as this entire subject will be thoroughly investigated in later chapters devoted to selection and preparation of the foods you plan to keep in your freezer. Refer to these chapters for step-by-step procedures.

You will want to know, however, what packaging materials and equipment to have on hand, ready to be put to immediate and efficient use, when you bring home that side or hindquarter of beef, the young spring lamb or suckling pig, the crate of fresh-killed chickens or the bulging creel of a day's fishing excursion.

Buy all your packaging materials well in advance. The most desirable types and brands may not always be available locally; there are too many kinds for any one source to be able to stock them all. Even if you discover a store or frozen food locker plant in your neighborhood which maintains a complete supply of the wrappings and containers you prefer to use, seasonal rush purchasing by other home freezer owners may deplete stocks just when you need them most.

Be forewarned—your store or locker plant will probably present a confusing array of papers and sheeting, all labeled "freezer wraps for meat." Side by side on the shelf you may find a roll of wax or butcher paper priced around 79 cents or less, whereas a roll of scientifically processed freezer paper may cost as much as $2.50. *Don't skimp when you buy packaging materials!* It is not necessary to spend a fortune, but it is foolhardy to select materials on the basis of price alone if what you get for your money is inadequate or unsatisfactory.

Remember, the costliest item in your freezer is meat. You may put in a year's supply of beef at a time when the price is seasonably low. Unless that prime roast or thick steak is as handsome and succulent as it was ten months ago, when you froze and stored it, you will have defeated your purpose of providing better meals for your family. If badly packaged pork chops removed after three or four months in the freezer have developed an off taste and must be discarded, you have been extravagant instead of economical.

There are two basic rules for selecting freezer packaging materials:

1. Never, under any circumstances, choose what is recognizably plain wax paper or ordinary butcher paper, no mat-

ter what its package label claims or how earnestly a salesman tells you it is suitable for freezing purposes. It is not.

2. Inasmuch as you cannot be presumed to be a paper expert, read the manufacturer's printed claims carefully and look for a statement which says, in effect: "This paper is moisture-vapor-proof and contains an oxygen barrier. It is pliable and strong."

The following wrapping materials for meats are recommended on the basis of personal trials and observation combined with the results of long-term experimentation made in impartial tests conducted by frozen food research laboratories at universities and government agricultural stations.

ALUMINUM FOIL

Look for and select brands of foil specified as *freezer weight.* Do not use lightweight foil for freezer storage, except for dry baked goods such as bread, rolls, unfrosted cakes and cookies.

While heavy aluminum foil meets most of the requirements for a meat wrap, it is somewhat costlier than other equally satisfactory materials. It certainly looks handsome in a freezer, if you are concerned with aesthetic effects.

Foil has one disadvantage, however, which is that no matter how careful you are in your original packaging or subsequent handling, tiny punctures frequently appear as the result of a number of causes: The package may be bumped up against the sharp corner of a rigid container in the freezer; a bone or sharp edge within the package may exert stress; you may handle the package while wearing a pronged ring; you may inadvertently drop it on the floor. Any number of things can happen, and when one of them does the food within the punctured foil is vulnerable to dehydration or oxidation or both, for vapors are able to pass genie-like through the tiniest of openings.

Because of this, I rarely use foil without a protective outer-wrap of heavy kraft paper or stockinette, a stretchy cotton mesh material sold in rolls (which of course conceals its attractive appearance), or without first wrapping the meat in a sheet of cellophane or plastic film (which makes the foil outer-wrap a needless extravagance).

For the most part, I use aluminum foil for wrapping very choice roasts or elegant cuts like filet mignon, principally for the pleasant feeling of luxury it gives me.

Heavy foil does have a plus value, however, that is not shared by any of the other meat wraps. Foods can be *cooked* in it, sealing the juices in and eliminating the use of a pan. (This should not be done with any of the meats or poultry you think of in terms of roasts, however, unless you leave the foil open at top; roasting is always done with dry heat.) This may appeal to many housewives. I am the kind of cook who likes to see what is going on, and so I rarely avail myself of this practical application of aluminum foil. There is one exception: When I have packaged leftover cooked foods in foil and frozen them, I may pop the package into a warming oven as is. I do this primarily to save time.

Although foil has a tendency to mold and cling to whatever it is wrapped over, it is wise to seal folded ends with one of the available freezer tapes which can then be marked with a grease pencil to identify the contents. If you use a stockinette over-wrap, you can insert a marked piece of paper between the foil and the protective stockinette for easy identification later.

CELLOPHANE

You may find several brands specified as *freezer weight*. Do not use the kind of cellophane for freezing which artists use for protecting their drawings, or the kind sold in stores as gift wraps. These are usually too light and brittle for extreme low temperatures.

Moisture-proof, heat-sealing, anchor-tight cellophane, sometimes designated on the label as "M.S.A.T.," is an extremely useful material for the freezer. It is easy to work with and, like foil, very good-looking. It creates a satisfactory barrier against both dehydration and oxidation, but becomes somewhat brittle under low refrigeration. For this reason, it is preferable to use an outer-wrap of heavy kraft paper or stockinette if the package is to be stored for a long period and is likely to be handled from time to time as other foods are placed in the freezer. The outer-wrap should be sealed with freezer tape, and the cellophane either taped or heat-sealed.

When using either aluminum foil or cellophane for bony meats or poultry, an additional precaution against fracturing is to cushion the protruding bones and the jutting surfaces of the poultry (legs, wings, etc.) with little wadded-up pieces of wax paper or cellophane before wrapping. Take pains to exclude all air within the package by pressing the foil or cellophane close to all sides of the food as you wrap.

THERMOPLASTIC FILMS

Plastic wrappings come in handy dispenser cutting-edge rolls and are wonderfully easy to use as meat and poultry wraps. They are undoubtedly the most pliable and moldable of any of the sheeting materials. (The term "thermoplastic" means that a basic chemical compound such as ethylene, styrene, ester, etc., has been transformed into plasticity by the application of heat.) The nature of this material is such that it can be *sealed* with heat, a use which will come under more pertinent consideration when we discuss the packaging of vegetables and fruits.

Most of the plastic films available in sheet and bag form combine the virtues of transparency or translucence, light weight, toughness, high resistance to punctures, tears or cracks, flexibility under even extremely low temperatures, freedom from odor, taste or toxicity, heat sealability and a very low rate of moisture-vapor transmission.

Because, however, some of the polyethylene products are not completely effective against oxygen transmission, I usually prefer to employ an outer-wrap of a known oxygen barrier when I package meats, fish or poultry for long-term storage.

When using any of the films for meat, poultry and fish, be sure to exclude all air and seal either with heat (a little clumsy to attempt on a meat package) or with one of the freezer tapes. In a pinch, which means that I was temporarily out of other sealers, I have used transparent tape to secure my packages. It seals all right, but just try to find where the tape ends and the film begins when you take a film-wrapped package out of the freezer.

Like cellophane, the transparent plastics provide a high degree of visibility. This is especially advantageous if, like me, you frequently pack a small quantity of some leftover food and tuck it away in a hurry. If you have forgotten or neglected to label the package, it's nice to know at a glance what that odd-looking little bundle is without having to unwrap it.

In my pioneer freezer days, for example, I had a habit of putting a lone croquette or a small piece of uneaten roast beef in a piece of aluminum foil, thinking, "I'll remember what *that* is." Months and many repetitions of this habit later, I would reach into the freezer and emerge with what I thought was a nice bit of sandwich meat for my lunch, only to discover when I unwrapped it that it was a couple of my mother's brownies, stored away for a ready treat when one of the neighborhood children dropped in to visit.

COATED (LINED) AND LAMINATED FREEZER PAPERS

There are a lot of these around, in many guises and under many brand names. Sometimes a local frozen food locker plant manager will have one of the large paper companies supply customer-sized rolls of the same paper he uses for his larger operations, and imprint the cartons with the name of his locker plant.

Good freezer paper is the cheapest material you can use for meat, poultry and fish because it can be used without an additional inner- or outer-wrap; *but be careful.* There is a great difference in the effectiveness of the various papers on the market.

Don't ever let anyone tell you that plain heavy kraft paper is all you need for wrapping meat to be put in the freezer. Heavy kraft is fine as an over-wrap for cellophane or foil packages, but all it will do is hold the inner wrappings more securely, prevent scuffing and tearing, and provide a space for writing down what the contents are. It offers no protection whatever against water-vapor or oxygen transmission, and therefore virtually invites dehydration and oxidation of your precious food.

Coated papers: As the name implies, these are kraft papers that are lacquered during manufacture with another substance, usually either wax or liquid plastic. They are somewhat less expensive than the lined or laminated types, and they offer a certain amount of protection in the freezer. For foods you do not plan to store over any great length of time, these are very satisfactory and their lower price is certainly a consideration.

When you use them, be sure to place the food on the coated side of the paper, with the untreated kraft surface facing the table on which you are working. You will be able to tell which is which by feeling both sides and by comparing their appearance. The coated side is generally slippery or waxy as well as shiny, whereas the untreated side is dull and can be written on easily with a plain lead pencil.

Some very good coated papers are put out in various lengths and widths by several of the country's large paper mills. These are labeled, simply, "freezer paper" or "freezer wrap," sometimes with an identifying trade name. The type of coating used in manufacture, whether wax or polyethylene, is specified on the carton. There are others as well, and when you buy them as a short-term freezing wrap be sure to look on the label for assurance that the kraft surface is a *wet-strength* grade.

Laminated papers: Lamination is a manufacturing process by which two or more sheeting materials are permanently affixed to each other by an adhesive or thermoplastic bonding agent to combine the properties of the component materials into one convenient packing material.

One very familiar application of the principle of lamination is the construction of plywood. A very thin ply of wood is glued under pressure to others. Each layer is placed so that the grain runs the opposite way of the next layer. This gives the finished plywood greater strength and greater molding qualities than any single ply of the same materials used.

If someone passed a law which stated that I, as a home freezer owner, would be allowed only one type of packaging material for my frozen meats, poultry and fish (instead of my good and jealously guarded American privilege of having as many different kinds as I can afford and want to play with), one of the laminated sheets would be my unequivocal choice. They are superb.

When home freezing began to show unmistakable signs of becoming a widespread national activity, scores of alert paper suppliers to the stationery, grocery bag, commercial packaging, frozen food and other industries jumped into production of papers intended for community locker plant and home freezer use. Paper-making is big business in this country, and a fascinating one.

Several of these companies, in their enthusiasm and zeal to get into a lively market as fast as possible, turned out experimental sheets which were entirely unsuited for long-term storage of foods at low temperatures. Unfortunately, this fact was not discovered soon enough to prevent wasteful food spoilage or disappointing results.

This chaotic situation has for the most part been rectified. Government agricultural stations, university extension services, home economists' careful tests and public indignation have seen to it that the paper companies leash their enthusiastic promotions long enough to do a little honest investigation of the qualities demanded for packaging materials.

Buying a roll of freezer paper is not like buying a box of stationery. If for any reason you are dissatisfied with the stationery—if, for example, it spreads the ink too much or is difficult to fold neatly or linty threads attach themselves to your pen, you can just stop using it and write a nasty letter to the manufacturer. You may even get your money back, along with an apology.

But when you buy freezer paper you are spending your money more for protection than for something which merely contains food. If the paper fails you, you may never even know it. All you know is that a quantity of valuable food has spoiled while it was in the freezer. Depending on your character, you may be inclined to blame either the freezer itself or your own shortcomings. *About 85 per cent of food failures in freezing is caused by inadequate packaging materials and methods.* The other 15 per cent is divided among trying to freeze unfreezable foods, poor quality of food, error in preparation and freezing too slowly.

You cannot give freezer paper a trial run as you can, say, with an automobile or a typewriter, before coming to a decision as to its desirability. By the time the paper has proved unsatisfactory, it is too late for decisions.

As much a part of every roll of laminated freezer paper as the wood pulp content which forms its base is the integrity and scrupulous purpose of its manufacturer. This is the hidden quantity, the X, of almost every commodity you buy; in freezer papers, it is of paramount importance.

PLASTIC AND LINED FREEZER BAGS OR TUBING

Although bags as such are mostly in the vegetable, fruit, dairy food and leftover departments, I have found so many excellent uses for them in packaging meats, fish and poultry that they should be mentioned here, as well.

For example, I use a small bag for chicken giblets before returning them to their original home, the cavity of the chicken I am freezing for roasting at a not too remote time. When I am ready to cook the bird, I have an important ingredient for the stuffing conveniently at hand.

I find the smaller sizes of either plastic or lined paper bags useful for small steaks and chops, variety and organ meats and even hamburger. For these I like double insurance, and so I am in the habit of gathering up all the little bags of one meat category and putting them in one big bag. This system has another advantage, too. It makes storage easier, and eliminates poking around in the freezer for a small package.

If reasonable care is taken in handling so as to avoid

puncturing, the plastic bags can be used again and again. They need only to be swished through warm, soapy water, rinsed thoroughly and allowed to dry. Test the plastic bags for punctures by filling them with water, rather than by trusting your eyes to detect holes.

The larger plastic bags are fine for poultry parts, halved or whole chickens, ducks and game birds. I would not suggest using a single bag for very long-term storage. I replenish my poultry supplies about three times a year, and the bags have never failed me. They are so easy to use that there is a great temptation to let them receive anything and everything you put in your freezer. Up to a point, you can succumb to this temptation. I even use them for enormous, company-sized steaks—after first wrapping the steaks in film, foil or laminated freezer paper. Putting them in the largest turkey-size bags is just my way of filing and maintaining inventory.

Wrapping It Up

MEAT, POULTRY OR FISH IN FOIL, FILM OR FREEZER PAPER

Two methods for making compact, airtight packages are most generally used for wrapping meat. These are the so-called "drug store wrap" and the "butcher's wrap." Of the two, I prefer the former. It seems easier to mold the paper around irregular surfaces with the drug store wrap, and the extra fold-overs hold the paper in close contact with the food when the package is turned this way and that as you wrap it. Both methods are illustrated and described on pages 106 and 107.

Whichever method you use, be sure to seal the package with gummed freezer tape or tie it tightly with strong string.

THE DRUG STORE WRAP

► Glossy side of coated or laminated freezer paper is the inside surface, which should face you as you start wrapping. Place meat in center.

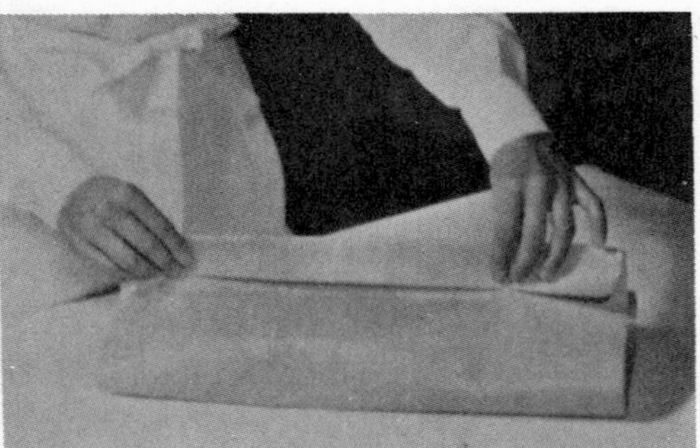

◄ Bring opposite sides of sheet together evenly, then fold down in series of folds until paper is tight against the meat.

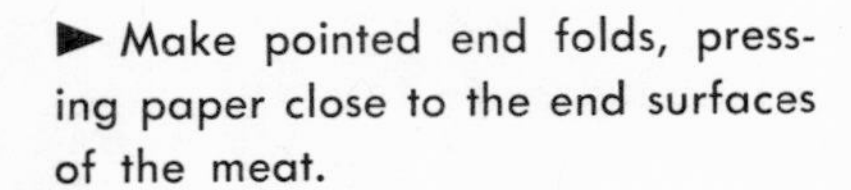

► Make pointed end folds, pressing paper close to the end surfaces of the meat.

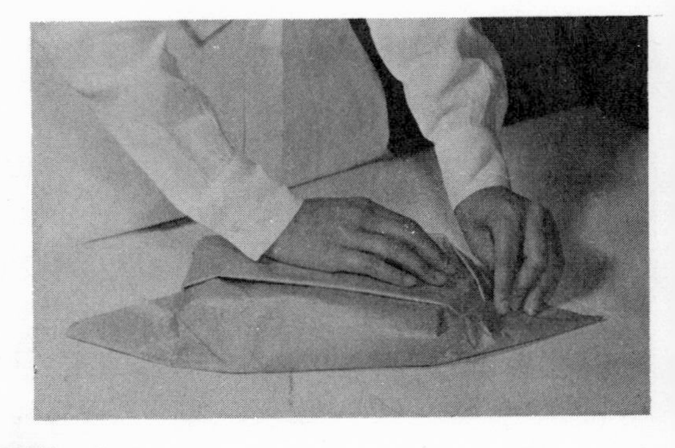

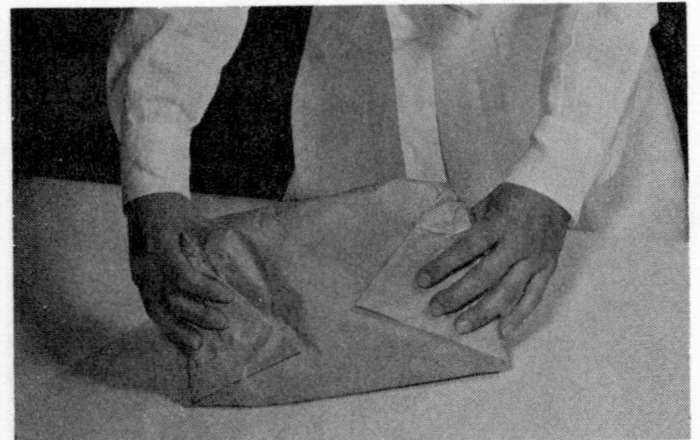

◄ Turn end folds under package. Seal with freezer tape or tie with string. Mark contents on outer surface with pencil or crayon.

Photos, Courtesy Marathon Corporation

THE BUTCHER'S WRAP

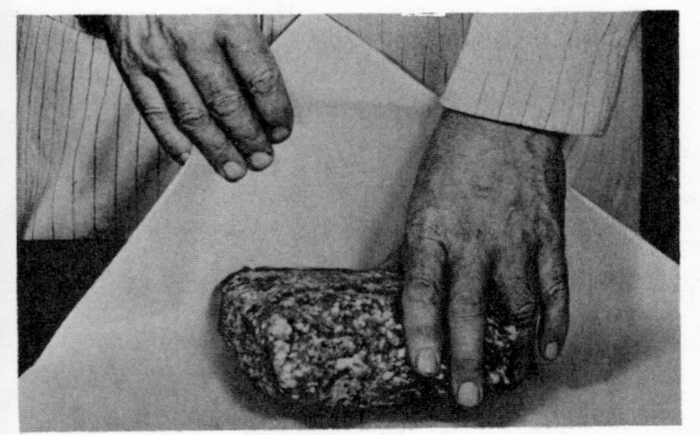

◄ Glossy side of coated or laminated freezer paper is the inside surface, which should face you as you start wrapping. Place meat on corner of paper.

► Bring corner over meat and roll toward opposite corner. Midway of roll, tuck the two side corners into center of package, pulling tightly against end surfaces of meat.

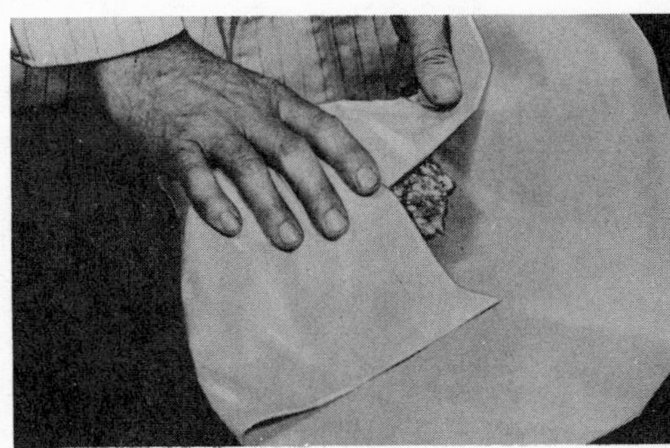

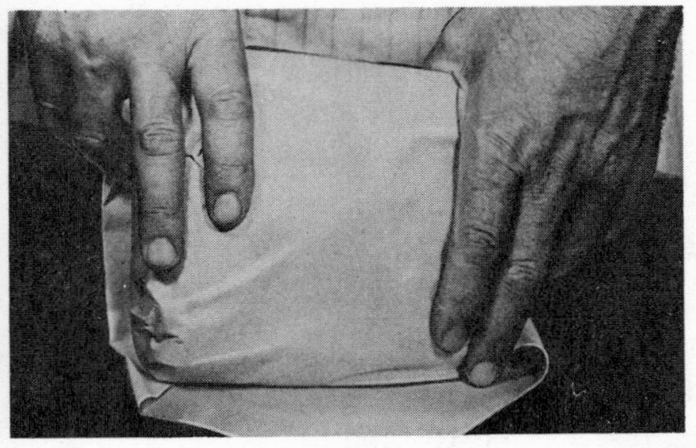

◄ Roll until paper is compact and tightly pressed against all surfaces of meat.

► Seal with freezer tape or tie with string. Mark contents on outer surface with pencil or crayon.

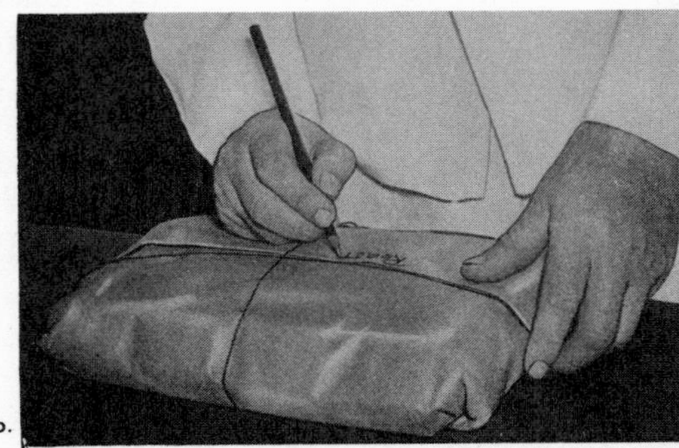

Photos, Courtesy Kalamazoo Vegetable Parchment Co.

Do not use ordinary rubber bands, as they will become brittle and break in the freezer.

By trial and error you will soon learn how much paper to tear off from the roll for each package. The tendency is to use too much. If you are working with a good laminated freezer paper, one thickness is all that is necessary; any more is wasteful. If you are using foil or one of the plastic papers or films, more than one thickness may trap some unwanted air between the layers. It is far better to use a single thickness of the material and protect the package with an outer-wrap of stockinette or heavy kraft paper.

To guide you, here are a few dimensions: Four average pork or lamb chops can be securely wrapped in a single sheet that measures 12 by 15 inches. A four-pound chuck roast requires no more than an 18- by 20-inch sheet, while 14 by 18 inches are sufficient for a three-pound pot roast. A five-pound hen will fit very compactly and snugly in a sheet that is 10 by 24 inches.

When wrapping meat, be sure to relate the size of the package to the size of your family or the meal for which it is intended. This may seem to be an injunction that is almost too elementary to be mentioned, but you'd be surprised how many times even a veteran home freezer enthusiast can make a silly mistake in this regard. To my everlasting chagrin, I once did something that I shall probably never be allowed to live down.

A gourmet friend presented me with a magnificent whole prime beef fillet, the kind you get a half-pound serving of in expensive restaurants on the à la carte menu.

"We'll have a feast," I promised my friend. "Next week—no, make it a week from next Wednesday—come to the dinner of your life."

I then popped the fillet into the freezer, for it was splendidly gift-packaged in cellophane with an outer wrapper of gleaming aluminum foil, which was handsomely and artistically criss-crossed with strips of bright red freezer tape. It was a perfect freezer package, and I complimented my friend for her thoughtful care.

The night before the scheduled dinner, I went to the freezer to remove the fillet, for I like my frozen meat to defrost slowly in the refrigerator before cooking it. Only then did I realize that what I had was a glamorous ten-pound lump of solidly frozen beef. It was unthinkable to defrost the entire fillet for a dinner party for three. Once defrosted, the remainder would have to be cooked at least partially before it could safely be returned to the freezer . . . and and what a culinary crime against the Very Best Beef *that* would be. The only way out of my dilemma was to cut off as much as was needed for tomorrow's dinner.

Did I say "cut"?

There is a handy little gadget, shaped like a coping saw, which is intended for cutting frozen foods. I have one. It works fine on some things. But the slender seven-inch-long blade was as ineffectual on that solid rock of meat as it would have been on a two-by-four oak beam.

In desperation, I fled to my work bench, came back armed with a two-foot wood saw and attacked the fillet. It was blasphemy and near massacre, but it worked. When I was finished, I had eight luxuriously thick steaks, five of which were properly wrapped separately and returned immediately to the freezer. I also had about half a pound of the most expensive sawdust I had ever seen.

At dinner the following night, my Siamese cats sat around watching us with a benign and expectant expression on their black masks. They knew darned well how good that meat was. It was several days before I could convince them that the experience was to be considered unique, and several more before they revived their interest in the ordinary muscles and organs which constitute their special treats. They left no doubt in my mind that they would willingly forego all else if they could round out their diet with sawdust.

So—when wrapping meat, do give a moment's thought to portion-control. (Detailed information about how to choose meat carcasses or gross portions thereof will be given in the chapters on procedures, along with instructions for separating them into table cuts.) It is a good idea, too, to slip-sheet small steaks and chops, chicken parts and pieces of fish that are packaged together. Just slip a double or folded piece of cellophane or plastic film between each piece. This makes handling easier later, when the food comes out of the freezer to be prepared for cooking. Don't attempt to use a single thickness for slip-sheeting, because it won't work.

After the food is wrapped, sealed or tied, *label the package.* Do this with a soft artist-type pencil or crayon, and mark the package with as much information as possible for your future guidance. For example:

One pot roast, top round, 4 lbs.—date frozen
One roasting hen, 6 pounds—date frozen
Two chicken breasts—date frozen
Four pork chops, loin, 2 pounds—date frozen
Six salmon steaks—date frozen (use within three months)

I frequently annotate my meat packages in a rather subjective way. The other day I took a package from my freezer on which I found the following intelligence:

"Butcher's tenderloin—2 lbs.—cubed for skewer. Save for Company, especially the Torreys. Kabobs?"

I always tell myself that this efficient system will justify my foresight when the freezer is loaded to the gunwales, and I am still confident that it will. But the day I took out that particular package was the day *after* the Torreys came to dinner and ate veal scallopini.

MEAT, POULTRY OR FISH IN PLASTIC BAGS

Handy as they are, plastic bags present two problems when used alone for meat, poultry and fish. Air must be excluded, and punctures must be avoided.

Excluding the air from a plastic bag containing an unevenly shaped solid mass is not as easy as it sounds, although it is possible to get most of it out.

One way (not one I especially care for, however) is to immerse the filled bag in a tub of water. The surrounding water exerts enough pressure to force the air out. While the bag is still immersed, and after the air is expelled, pinch the bag together tightly at the point closest to the food. Holding it pinched like this, raise the bag and twist its top quickly around and around, like making a little girl's curl. Then bend the ropy twist back on itself, forming a "goose neck." Seal it tightly with several turns of a coated wire twister.

Another, easier method is to hug the bag and its contents to you and start pressing it with your hands from the bottom up. Make a gooseneck closure and fasten with a twister.

When bagging whole fowl, truss the wings and legs to the body with string in order to make a more compact and easily handled package. Whole fowl should have the sharp, protruding edges cushioned with cellophane or paper before being put into the bags, and chickens cut into parts should be individually wrapped first in cellophane or plastic film for easy separation when you remove them from the freezer.

Fish, too, should be wrapped in an extra sheet of heavily waxed paper or cellophane before being packaged in a plastic bag. This additional precaution is to keep odors and moisture from escaping into the rest of the freezer. Actually, I never use plastic bags alone for fish. I prefer either to put the fish in a bag and then into an oblong moisture-vapor-proof container of the type used for vegetables, or to wrap it in waxed paper and then in the best laminated sheet I have in the house. I take no chances on seepage or rancidity.

Containers for Vegetables and Fruits

In the not-so-distant past, when freezers were in far fewer homes than the millions they bless today, there was comparatively little choice in containers. They were so scarce, as a matter of fact, that I used to flatten and hoard the boxes in which my store-bought frozen foods came packaged. If they were good enough for Birds Eye or Snow Crop, I figured, they were good enough for me.

Since the first edition of this book, home freezing has become so much a part of domestic American life that industry now spends a great deal of time, effort and money on freezer packaging materials research, to the end that products available to freezer families are constantly being improved and increased.

In the wide selection available you will find oblong, square and round paperboard containers with permanent or removable liners of glassine, plastic, cellophane or wax-impregnated paper.

There are waxy cartons with cardboard or plastic lids. There are rigid plastic boxes with flexible lids, and flexible plastic boxes with rigid lids.

There are special glass jars without shoulders, tapered so that their openings are wider than their bases for easy emptying. There are aluminum boxes, some of them tinted with color for food identification. There are aluminum foil containers that can be baked or cooked in, frozen and later heated in the oven. There are handsome stainless steel and ceramic containers.

In addition to the containers offered for sale, moreover, there are some "free" ones available to purchasers of commercially frozen heat-and-serve foods packaged in various sizes of heavy aluminum foil boxes. To be found in frozen-food cabinets, along with the prepared foods they contain, these sturdy boxes can be washed after their original use and saved for future freezing purposes. The larger of them, I've discovered, are excellent for precooked casseroles, while the smaller ones do service as containers for leftover sauces, gravies and portions of main dishes too small for a meal but large enough for a snack.

When re-using these foil boxes, be sure to cover the open tops securely before you entrust the containers to the freezer. Depending on the contents, I protect mine with foil or plastic film or pop them into polyethylene bags. Removed from the freezer and uncovered, they can go directly into a heating oven.

PLASTIC CONTAINERS

These are things of beauty and joys forever. In my mind they represent the happy state of affairs which comes to pass when the freezer owners' requirements are anticipated and met by manufacturers' product planning.

For fruits, juices, vegetables and some leftovers I prefer them to bags, boxes, cartons, tubs or jars, and their understandably higher price is an investment in convenience and ultimate economy. After all, you can use them practically forever with reasonable care that they do not break. Although plastic, one of the wonders of our modern era, is pretty sturdy stuff, very low temperatures are likely to make it brittle and subject to stress.

You will find many variations of the plastic box in your neighborhood, and the probabilities are that they are all good, so long as they are identified as *freezer* boxes. The thing to watch for is the type of lid supplied with the box, for to be suitable as freezer containers the lids must fit snugly to form a vacuum-tight closure. Lids that merely rest on the tops of the boxes are not sufficient. They should be the flexible snap-on type with a slight overhang which presses over the container.

A helpful characteristic of all rigid plastic boxes is their total visibility. The frosted strips are sometimes redundant, when you have put away such readily identifiable foods as peas, beets or carrots. However, it is smart not to depend too much on recognizing what's inside. It is difficult to distinguish between a box of frenched green beans and cut pods, and, when solidly frozen, strawberries are easily mistaken for raspberries. Better take the few extra seconds and record the contents.

More and more manufacturers in the freezer supply business are turning to a wax-like plastic, polyethylene, for their boxes. This is the same basic material with which you are familiar in plastic bags, but it is processed into a different form.

OTHER CONTAINERS

Many food processors' packaging ideas provide very nice reusable containers of plastic or rigid foil in all sizes and shapes.

Milk cartons, also, can be converted into freezer containers by cutting off the tops. Be sure to wash all of these carefully and be sure the water you use is not so hot that it melts the paraffin.

Another money-saving idea leading to a stockpile of usable freezer containers is to save empty vacuum coffee cans and the variety of glass jars in which mayonnaise and peanut butter are occasionally packed. The kind of jar I mean has straight sides, no shoulders. Unless you have the time and patience to wait around for a complete defrosting of the contained food, don't ever use an ordinary shouldered jar for freezing. You probably won't more than once, but sometimes once can be too much.

An economy-minded friend once telephoned me gleefully to say she had just prepared and frozen two dozen quart pickle jars of chili con carne, a great favorite with her Canasta club. She telephoned a few weeks later to say that the club had convened at her apartment the night before for cards and chili, but had to be satisfied with cards and scrambled eggs.

She had forgotten to take the jars out of the freezer until a little while before her guests were expected, and even after several hours of heated card playing the chili was still solidly frozen and could not be coaxed through the narrow jar opening.

If you save and use coffee tins, be sure first that they are made of rust-proof metal and, second, that they are thoroughly washed and deodorized before you enlist them as containers. Coffee oil has a way of clinging. Such tins, lined with wax paper or cellophane, make very acceptable containers for leftover foods and those little bite-size cookies.

If you are a homemade plum-pudding enthusiast, coffee tins, the pound size, are just right for steaming or baking the puddings according to your favorite recipe. When the tins and their contents are cooled at room temperature, they may be put in the freezer after you seal the lids with freezer tape.

Packaging Vegetables and Fruits

There is no great trick to this, and the process is not nearly as fussy or worrisome as meat-packaging. Plastic bags used alone may be a little floppy and sloppy, but you soon develop amazing dexterity and inventiveness. I have used them happily for fresh chicken livers, and you know how wiggly they are. For this I created a "shovel" out of fairly stiff kraft paper, which I inserted well inside the bag, leaving about eight or 10 inches of my makeshift shovel protruding. I piled the livers on the kraft paper, then raised the bag and paper together in a graceful, scooping motion. Holding the bag upright, I slid the paper out—and voilà!

Another trick I use when packaging cut vegetables or fruit sections in polyethylene bags is to spoon the food into a tall glass or jar, put the opening of the bag over the top of the glass like a nightcap, and quickly reverse their positions.

Rigid cartons or plastic boxes present no problems at all, and the bag-and-box combinations usually come equipped with a cardboard funnel. *When using rigid containers, be sure to follow directions for leaving expansion-space at top, otherwise your containers will split when the food expands during the freezing process.*

Heat-sealing of cellophane and plastic or plastic-lined bags is easily accomplished with a warm iron, a marcel curler or with one of the special gadgets available for this express purpose from many hardware, chain, and mail-order stores.

Well, there you have it. Or them. The big, wide, wonderful variety of wrapping and packaging materials ready for you to choose from. Which ones should you buy? How much should you buy? Because home freezer capacities vary from four cubic feet up to thirty feet, and because each family's preferences in food are different, there can be no all-inclusive answer to these questions. I will venture these tentative recommendations for a basic freezer material wardrobe, while pointing out that there is great flexibility of choice:

Recommended Basic Packaging Materials to Keep on Hand

Laminated Freezer Paper—one 100-foot roll, 18 inches wide (sufficient for a cut-up hindquarter of beef, about 250 pounds).

Cellophane—one roll for special wrappings of delicate foods such as variety and organ meats, and for slip-sheeting steaks, chops, chicken parts, fish fillets, etc. (Heavy waxed paper or plastic sheeting may be substituted as slip-sheets. Also, used polyethylene bags which have seen their service in freezer duty and are punctured or ripped may be cut apart and saved for this purpose.)

Polyethylene, Vinyl or Pliofilm Bags—a trial package of each of every available shape and size will not constitute a tremendous investment and will be more useful than you may suspect. You will find your own preferences among the various sizes for future guidance. The chicken and turkey sizes are most frequently used, and for many foods besides chickens and turkeys. They are excellent for storing bread, rolls, cakes, pies, etc.

Bag-and-Box Combinations, Paper Tubs or Plastic Containers for Fruits and Vegetables: Start with a dozen pints and quarts each of the bags-and-boxes, the same number of tubs, and get a few of the plastic containers. Try all of them. If budgeting is necessary, you will find that the bag-and-box combinations are least expensive. If you are going all out, then by all means splurge on the plastic containers.

If freezer space is at a premium, remember that square or oblong packages take up less room than round ones. Keep in mind also that when you buy a bushel of peas, you are going to need from 12 to 15 pint containers in which to package and freeze them. Approximate yields of all vegetables and fruits, in terms of pint containers, are given throughout Chapters 16 and 17.

Heavy Aluminum Foil—one roll, for fancy packages.

Gummed Freezer Tape—one roll.

Stockinette—one roll, if you are going to package meats and poultry in cellophane.

Equipment and Gadgets—Musts: Sharp kitchen knives,

scissors, cutting board, a very large enamel or aluminum pot for blanching, preferably one with a colander-like insert for easy removal of vegetables from the blanching water.

Optional, but nice to have, are a small frozen food saw, a heat sealer (your own electric iron will do), a carton-filler device if you are very gadget-minded, otherwise work out your own destiny.

Some Packaging Do's and Don'ts

Do buy your materials well in advance of the days on which you have scheduled your freezing activities.

Do keep an inventory record of the foods you freeze. A loose-leaf notebook will serve for this purpose.

Do be sure to expel all air from packages containing meat, fish and poultry.

Do use either the drug store or butcher's wrap when packaging meats.

Do use materials which are moisture-vapor-proof and barriers against the transmission of oxygen.

Do seal all packages, either by heat or with freezer tape.

Do leave expansion room in containers holding wet-packed fruits and vegetables.

Do package all foods in meal-size portions.

Do label all packages fully, and date them.

Do keep packaging materials in a cool place, where they won't dry out. The freezer itself is a fine storage place, if you have room.

Don't freeze any food in plain waxed paper or in butcher's paper alone.

Don't put very hot foods in waxed tub containers.

Don't apply a hot iron directly to cellophane or plastic films.

Don't use quart containers if your family eats only a pint at a time.

Don't re-use broken or punctured containers.

Don't wash plastic containers in very hot water.

6. Mostly About Meat

Meat is the main course of our national diet. Excepting only vegetarians, most American homemakers, restaurateurs and dieticians plan their menus around meat, which is the most expensive single item in food budgets. Well-to-do families take for granted frequent servings of the choicest table cuts at home and when dining out; less financially privileged folks are ingenious about ways and means of preparing the equally nutritious so-called "economy" meats. From one income bracket extreme to the other, and in between, when Americans ask, "What are we having for dinner tonight?" they invariably mean, "What's the meat course?"

Meat is not only preferred by us for its good taste and satisfying qualities, it also makes an outstanding contribution to our diet according to high nutritional standards. It is a very concentrated source of high quality proteins, minerals, extractives and important factors of the Vitamin B complex. At the same time, it is comparatively low in calories.

Freezer families, no matter what their income bracket may be, find that the quality, palatability and variety of their meals improve to a marked degree as they learn to use their home freezers as well-organized storehouses for fresh meat. They are no longer dependent on availability in local markets of the cuts they like. They are not threatened by periodic shortages caused by national or local emergencies such as war, strikes or cattle-destroying storms. They are not obliged to pay premium prices during inflationary times.

Actually, they can, through proper planning, save so much money and yet eat so much better than non-freezer owners whose meat costs are often twice or three times as much, that the latter raise incredulous eyebrows when they are told in all seriousness that a family of four hungry people can eat sizeable portions of *meat every day*, if they wish, for less money than a non-freezer owner spends during a year which of economic necessity includes scores of meatless days and meat-substitute dishes.

Freezing is the only present method of preserving the original flavor, texture and nutritive value of meat—if proper procedures are followed. We know now that fast-freezing is required (Chapter 4) and protective packaging imperative (Chapter 5). It is of primary importance to realize, moreover, that no home freezer is endowed with magical or alchemistic powers; the meat you take out of the freezer will be no better than the meat you put into it. Always remember, too, that you cannot overload your freezer and expect it to freeze the excessive load properly (Chapter 4).

The first rule for freezing meat products is to select sound, high-quality grades from healthy, well-conditioned animals. But how are you going to know what the age, home life and health of an animal was during its lifetime when you see trimmed pieces of it displayed in a butcher's case? Steaks don't come supplied with biographies.

Because I believe that you, as a home freezer, are interested in learning about the sources of the foods you preserve, several of the chapters to follow will attempt to explain how you can easily classify the various meat products as well as how to buy them in bulk to save money and how to process them for freezing.

Preparing Meat for the Freezer

If you decide to avail yourself of the services and facilities of a local frozen food locker plant operator or of a cooperative butcher with a quick-freezing cabinet, he will follow your instructions as to the size and number of packages you want your meat purchase to provide. Furthermore, he will, for a nominal charge, relieve you of all preparation chores by cutting, trimming, wrapping, quick-freezing and labeling the carcass, side, quarter or wholesale cut. All you will have to do, besides paying the man, is to carry the frozen packages home and arrange them conveniently in your freezer after entering them in your log book.

It is more economical, very little trouble and extremely satisfying to package and freeze the meat yourself, keeping always in mind that, ordinarily, *no more than four pounds per cubic foot of freezer space should be frozen at one time* for best results.

If, for example, yours is a 10-cubic-foot chest freezer, plan to freeze no more than 40 pounds one day, 40 the next, and so on. Keep the meat in the refrigerator or in a cold room until you are ready to wrap it for freezing.

When your quantity purchase includes an assortment of fresh meats—beef, veal, pork, lamb and organs—start your freezing preparations with the organ meats first, pork next, then veal and lamb. Do the beef last, for it keeps very well under refrigeration and often improves in flavor and texture when allowed to age for a few days.

After the meat has been cut or divided into meal-size portions, wipe each piece gently with a dampened cloth and wrap individually in one of the materials recommended in Chapter 5 (laminated freezer paper is my preference for meat packaging). Refer to the illustrations pages 106 and 107 for the best and easiest methods of wrapping meat. Seal or tie each package securely and label it accurately with a soft pencil or crayon. Examples:

One four-pound rolled roast beef, choice
Two pounds of boneless lamb stew, neck
One pound of thin veal cutlets, leg
Four 1-inch pork chops, loin
One pound beef liver slices

Date all packages and enter them on your log sheet or in your inventory book.

Individually wrapped steaks and chops and packages of stew or ground meat may be gathered together and placed in large polyethylene bags for easy storing and greater accessibility. When you do this, put a slip of paper in with the packages and total them. Whenever you remove one or more packages from the bag, deduct the number from the total score.

A convenient way to package hamburgers or other ground meat patties is "by the yard." Do this by placing a row of equal-sized, flattened patties on a strip of cellophane about 6 by 36 inches, leaving about two inches of space between each patty. Cover the strip with another piece of cellophane of the same dimensions and fold the margins in close to the meat. Make a compact package by folding the covered meat patties back over each other accordion style. Put this cellophane package in a polyethylene bag, sealing air out with the goose-neck closure.

When you want hamburgers for dinner, you can merely

cut off as many of the patties as you need with a pair of scissors without disturbing the remainder. Pre-shaped hamburgers take less time to defrost than the same amount of ground meat packaged in bulk.

Ground meat of any type should not be salted before freezing, as salt accelerates the development of rancidity during storage because of the meat's greater exposed surfaces. You may, however, add your favorite herbs and other seasonings to ground pork or sausage meat before freezing it. Be especially careful when packaging ground meat to use a good oxygen-barrier material, for it is more vulnerable to oxidation than whole cuts are.

When you place meat in your freezer, be sure that the packages make contact with the fast-freezing surfaces and allow air circulation space around the packages. In chest style freezers, meat should be placed against the side walls or in a fast-freezing compartment, if you have one. In upright freezers, packages should be placed directly in contact with the freezer plate shelves, or against the side walls if yours is the wrap-around type.

Allow meat to freeze solidly before stacking it in the freezer and putting in additional quantities. Heavy, bulky roasts take longer to freeze through than do steaks and chops.

The foregoing recommendations are approximate maximums. If properly wrapped, fast-frozen and stored at temperatures which never exceed 0°F., beef and lamb of good original quality have been preserved without flavor loss for as long as two years, pork and veal at the same temperature and under the same conditions for as long as a year.

There is not much point, however, in keeping meat in storage for such lengths of time. The most desirable way to use your home freezer is to keep "turning over" its contents, putting away the various foods in sufficient quantities for a three- to six-month period at the seasons of the year when

RECOMMENDED MAXIMUM STORAGE PERIODS FOR MEATS

Type of Meat	*Months at 0°F.*	*Months at 5°F.*
Beef		
Roasts, steaks	12-14	8-10
Cubed, small pieces	10-12	6-3
Ground	8	4
Veal		
Roasts, chops	10-12	6-8
Thin cutlets, cubes	8-10	4-6
Ground	6	3
Lamb		
Roasts, chops	12-14	8-10
Cubes	10-12	6
Ground	8	4
Pork		
Roasts, chops	6-12	3-5
Ground, sausage	4	2
Pork or ham, smoked	5-7	3
Bacon	3	1
Variety meats:		
Beef or lamb liver, heart	4	2
Veal liver, heart	3	1
Pork liver, heart	2	1
Tongue	4	2
Kidneys	3	1
Sweetbreads	1	—
Brains	1	—
Oxtails	4	2
Tripe	1	—
Spiced sausage or delicatessen meats	2-3	1

they are most plentiful and eating those same foods fresh while they are in season. In this way, you will have a supply on hand when the foods are out of season, scarce and more expensive.

WHEN TO BUY MEAT IN QUANTITY FOR FREEZING

Your geographical location and the amount of empty space in your freezer will usually determine the exact time of the year when you order large supplies of your family's favorite meats for long-term storage. The market fluctuates from year to year. However, the following general guide, coaxed from the United States Department of Agriculture, is a composite of best supply and lowest prices on a national basis, computed over the past several years:

Beef	U.S. Prime, Choice and Good	From March to June
	U.S. Good and Commercial	October and November
Lamb	Hothouse and Genuine Spring	March and April
	Spring and Yearling	July through January
Veal	Youngest, most desirable	April and May
Pork		November and December; March and April

Descriptions of the various U.S. gradings will be found in the chapters dealing separately with each type of meat.

How to Make and Freeze Soup Stock

Whether you do your own butchering or have it done by a frozen food locker plant, there will always be a quantity of good bones and meaty trimmings that go along with your bulk meat purchases. Unless you have a tremendous amount of unused room in your freezer, these take up too much valuable space. They are too rich in extractives and nourishment to be thrown away, however, and you can preserve their nutritive value by freezing them in the form of concentrated stock for soups or gravies.

Crack or saw beef or veal bones, or a combination of the two, and cover them with water. Bring slowly to a boil and simmer for two or three hours over low heat. (Instead of plain water, you may for additional food value use liquid in which vegetables have been blanched or cooked.)

Let the unseasoned liquid cook away until a strong broth remains. Strain and allow it to cool.

Pour the stock into refrigerator ice-cube trays or into small-sized tub type or plastic containers. Freeze at once. Ice cubes may be popped out of their trays and stored in polyethylene bags.

If you want to freeze larger quantities of stock for multiple recipes, freeze it in bread tins or even in oblong cardboard candy boxes which have been lined with heavy waxed paper or cellophane. When the blocks are thoroughly frozen, remove them, package in laminated freezer paper and store away the "bricks." This method conserves freezer space and does not tie up your supply of containers.

BULLETINS

The following pamphlets and bulletins are available:

From the United States Department of Agriculture, Washington, D. C.:

#93—Freezing Meat and Fish in the Home (20¢)
#G58—Shopper's Guide to U.S. Grade for Food

From the Institute of Meat Packing, Chicago, Ill. (free);

Beef, Lamb and Veal Operations
Pork Operations

From the University of Illinois, College of Agriculture, Chicago, Ill.

Beef for the Table (25¢)

7. Beef . . . from the Barn to Your Freezer

On the hoof, beef animals are divided into five classifications according to their sex, age and parental status:

A *steer* is a young male which was castrated in calfhood.

A *stag* is an older male which was castrated after being bred.

A *bull* is a male used principally for breeding purposes.

A *heifer* is a young female which has never been bred.

A *cow* is a mature female which has borne one or more calves.

Before home freezing came into the prominence it now enjoys, a housewife who bought her meat at the local butcher shop could identify at least approximately the kind of beef animal from which a certain roast or steak was cut. She recognized the federal stamps which grade all meat inspected by the United States Department of Agriculture, Food Distribution Administration. A round purple stamp about the size of a quarter is marked on the fatty surface of the beef, signifying that the meat has been approved as meeting the U.S. standards for healthy animals, slaughtering regulations and sanitary conditions. This is the federal inspection stamp.

The knowing housewife looked also for another purple stamp which, allowed to remain on the more expensive and desirable cuts, identified it as "Prime," "Choice" or "Good." This is the federal grading stamp.

Such federal inspection and grading are required by law for all meat shipped from one state to another, and until very recently somewhat more than 75 per cent of all meat consumed in American homes, restaurants and institutions traveled over state boundaries. Meat not shipped interstate is generally inspected and graded according to legislative regulations adopted locally by states, counties or cities.

Government-inspected beef is stamped as to its *grade* according to the age and sex of the original carcass, and the grading depends upon the *conformation, finish* and *quality* of the beef, all three words having very definite meanings in beef language.

Conformation refers to the general physical appearance of the carcass, side, hindquarter or wholesale cut from which table cuts are later sliced. To be adjudged good, the conformation of a beef carcass must represent a stocky animal with a short neck and shanks, full loins, plump rounds, well-fleshed ribs, thick flanks and shoulders.

Finish relates to the color, quantity, type and distribution of fat. Good finish means a smooth, even covering of fat over most of the outer surface, and liberal deposits between the large muscles and muscle fibers.

Quality signifies the firmness and strength of muscle fibers, the intermingling of fat with lean portions (called "marbling"), color and texture. In order to merit description as good quality, a cut of meat must be well-marbled and of fine grain, and the color must be typical of the particular meat.

Aging of Beef

After slaughter, a carcass must be cooled quickly at temperatures of from 34° to 36°F., for it is difficult to prevent souring around the bone in heavy animals if the temperature rises to 40°F. or higher. The chilling process must be carefully controlled within the indicated temperatures, for the carcass at this time should be protected against freezing.

Beef improves if it is aged or held under controlled refrigeration from six to ten days before being cut into its component parts. If it is held for a period of time up to thirty days, it is even better. This is what is called "aged beef," a premium commodity that commands premium prices because the aging process is expensive. The refrigeration itself is costly in terms of electric power, and there is, moreover, a considerable weight loss due to shrinkage during the aging process.

Slow-aged beef looks a little different from the steaks and roasts we are most accustomed to seeing in butcher cases. It is darker than the fresher beef, sometimes almost purplish in color.

You may from time to time see beef which is advertised as "aged" which does not look quite so dark. This is because the meat industry has developed a new method for "aging" beef rapidly by means of ultra-violet rays and higher temperatures. This example of tribute to the modern god of speed is, in a way, similar to American winegrowers' habit of aging their vintages artificially instead of in the European tradition which permits time to do it naturally.

Government Grading Standards of Beef

New grade classifications, adopted by the United States Department of Agriculture in 1952, are now standard for all cuts of the beef you buy. Grades are determined by government inspectors who base their decisions on their observations after carefully examining a carcass and checking its characteristics of conformation, finish and quality.

U.S. Prime Beef is without exception from steers and heifers only. To earn a federal inspector's stamp which designates it as "prime," the carcass' conformation, finish and quality must be top-notch, meeting the following descriptions:

Conformation—compact, blocky and thickly fleshed; loins and ribs are full; rounds are plump and extend to the hock joints; neck and shanks are short.

Finish—smooth, uniform fat covers the entire carcass; pelvic cavity and kidney are heavily larded with fat, as are the spaces between the backbone (called "chine") and the ribs. Fat and lean are closely intermingled between the ribs. This pattern is called "feathering." The fat throughout is firm, brittle and waxy.

Quality—lean portions are exceptionally firm, smooth and velvety; color is uniform and bright, ranging from pinkish red to deep blood-red, depending on how long the carcass has been aged; marbling is extensive and uniform; backbone and brisket bones (the brisket is the part directly in back of the animal's foreshank) are somewhat soft and their color is red, while the cartilages are pearly white.

While a very large proportion of the government-graded U.S. Prime beef is sold directly to swanky hotels and restaurants, a considerable amount nevertheless does find its way to butcher shops and locker plants where you may encounter it. If, as a home freezer owner, you are shown a quarter or a cut which is represented to be U.S. Prime beef, look first for the federal inspection stamp and the grading stamp which are printed on exterior fat in purple ink. No beef is "U.S. Prime Grade" unless it is so stamped.

You may, however, run across a nearby rancher who, not shipping his meat out of the state, comes under the jurisdiction of local authorities. If you learn to recognize the foregoing invariable characteristics of the prime grade, you may be able to treat yourself and your family to this delicious, top-quality beef at tremendous savings. Pay especial attention to the larger bones, for by them you can determine the age of the original animal. If the chine bone (backbone) is pink or red and seems to be porous, it is a young steer or heifer. White, hard bones are characteristic of older animals.

U.S. Choice Beef comes from steers, heifers and young cows which have been raised as beef cattle rather than for breeding or milk.

Conformation—moderately compact, blocky and thickly fleshed; loins and ribs are slightly flat; rounds taper toward the shanks; neck and shanks are moderately short.

Finish—surface fat is moderately patchy, but interior fat found in the pelvic cavity and around the kidney is plentiful; there is a slight extension of fat between the chine bones and inside the ribs; feathering is limited, and all fat is brittle and either creamy white or tinged with yellow.

Quality—lean portions are moderately firm, smooth and velvety; color is usually bright and may be somewhat two-toned or appear shaded, varying from light to dark red; marbling varies from slight to extensive. Depending on the age of the animal, bones are either soft and red or relatively hard and whitish.

U.S. Good Beef, from comparatively young steers, heifers and cows, is the grade most frequently found in popular markets and chain stores, although an occasional shipment of "Prime" or "Choice" beef may be advertised during or soon after the peak of the slaughtering season.

Conformation—slightly compact, blocky and thickly fleshed; loins, ribs and rounds are only slightly full; neck and foreshanks are somewhat long and thin.

Finish—determined by the animal's age. The surface fat of younger carcasses is thin, that of older animals thicker; fat between chine bones and inside ribs is slight; feathering is variable from abundant to slight, and all fat is soft or oily.

Quality—lean portions vary in firmness and texture from smooth to not-so-smooth, from fine to almost-coarse, depending on the age of the carcass; marbling is meager or moderate; color may be light red or fairly dark or two-toned; bones will be reddish or white, according to the animal's age.

U.S. Commercial Beef, from steers, heifers and cows, is frequently found in the retail markets, sometimes frankly as itself, sometimes represented as a better grade. During times of meat shortages or rigid controls, U.S. Commercial beef is often purchased in carcass, side, quarter, or wholesale cuts by local butchers or locker plants which do not have easy access to the more popularly desirable better grades.

The honorable vendor retains the purple government grading stamp, the less scrupulous removes it by shaving the fat where it appears and passing the beef off to the public as "Good" or "Choice." This is, of course, strictly illegal, but policing is difficult. The purpose in falsely up-grading the meat is a matter of profiteering; for, sadly, dishonest businessmen are with us always, and they can command higher prices and garner greater profits if the beef is presented to the innocent purchaser as a better grade.

There is nothing whatever wrong with Commercial grade beef. It is no less healthful and nutritious than the Prime,

Choice and Good standards; it is, because of its less generous coating and marbling of fat, inclined to be a little tougher and requires longer cooking. For some recipes, as a matter of fact, I actually prefer the Commercial grade and generally include some cuts of this meat in my wholesale purchases. I especially like the top round for pot roasts, the bottom and eye round for Swiss steaks, and have thoroughly enjoyed Commercial grade sirloin and short loin steaks. They are tasty, lean and quite a bit more economical.

Conformation—somewhat rangy, angular and irregular; thinly fleshed; flat, lean loins and ribs; long, flat, tapering rounds; necks and shanks have a tendency to be long and thin.

Finish—exterior fat may be thin and does not extend over the round or chuck; or, it may be moderately thick but unevenly distributed; little or no fat between chine bones and inside ribs, although some carcasses in this class may have fairly heavy development of fat in these areas; no feathering between ribs; small to large deposits of fat in pelvic cavity and over kidney; fat is soft, oily and rather definitely tinged with yellow.

Quality—lean portions may vary from firm to coarse and watery; little marbling; color is generally two-toned, shading from light to dark red; bones in some animals of this grade may be soft and red, while in others they are hard and white.

Other U.S. Inspected Grades are *Utility, Cutter and Canner,* considerably less tender but nutritious and wholesome. These rarely reach the retail level, being sold directly to meat packers for canning, sausage-making and meat product specialties.

Where and How to Buy Beef in Quantity for Freezing

Retail or wholesale butchers or community locker plants and nearby cattle ranches or farms with slaughtering, chilling and aging facilities are all good sources for quantity purchases of beef, and considerable money can be saved if you go about your buying cleverly.

Unless you are very sure about a neighboring rancher or farmer, it is best not to choose your meat on the hoof. Looking over a picturesque herd grazing contentedly in a pasture, you may be moved to point enthusiastically in the direction of a pretty brown bovine with large soft eyes and a philosophical expression and say, "I'll take that one!" The possibilities are that you will have chosen an animal too old to be tender, but the wicked farmer will agree whole-heartedly with you that that is a mighty fine piece of eatin' beef.

To be on the safe side, unless you are or know an expert at judging cattle, wait until a carcass is dressed and hung before making your selection, and then only after it has passed under inspection by an authority—either federal, state, county or city.

If it has been graded by a federal inspector, remember, it will bear the purple government stamps attesting to its health and classifying it according to one of the U.S. standards. State, county and city authorities have their own grading stamps, and it would be well to familiarize yourself with these in your locality. A letter to your state or city department of agriculture or board of health, or to your county agent, will bring you the desired information.

If, however, you are lucky enough to live in an area which is famous for its cattle-raising and feel secure enough to buy a side or quarter from someone you know, you can learn to recognize in advance the quality and palatability of the beef by comparing the carcass or cut with the characteristics described in the foregoing discussion of U.S. Government grading. They are specific and uniform. No one will be able to delude you into accepting his word for it that a cut is choice, for example, if you see with your own eyes that the bones are white and hard, indicating an older animal, and that surface and interior fat are yellow and meager, signifying a lower grade.

How to Cut Beef for Freezing

If you have great courage and confidence and are willing to profit by experience, you and your family can butcher a side or quarter of beef yourselves by studying the chart on page 137 and arming yourselves with a set of butchering implements. A great many freezer families do, and they have my open-mouthed admiration. Although I consider myself a rather daring creature when it comes to freezing, I draw the line at such earthy pioneer stuff, preferring to let a friendly butcher separate an animal into its edible portions according to my specifications. I find such supervision necessary, for unless I stand at his side repeating, "Thicker! Thicker!" he seldom slices steaks as generously thick as I like them.

If I lived closer to a community locker plant, I would avail myself of the facilities there; and if you do, that is where I would advise you to go, for the nominal cost of chilling and cutting and even, if you like, grinding, packaging and freezing, is not only worth what it saves you in time and effort, it also may make the difference between economy and waste. Unless you are pretty sure you know what you are doing, you may forfeit a lot of good beef cuts by not separating the muscles correctly.

You can, for the fun and experience of it, try your hand at separating one of the *wholesale cuts* (see chart, page 137) into its component steaks and/or roasts. This is not too difficult, and it is very rewarding both for the sake of economy and for the fine feeling of accomplishment it engenders. I frequently buy a complete loin or round and spend a gory afternoon making luxurious steaks and a battery of roasts. I

BEEF CHART

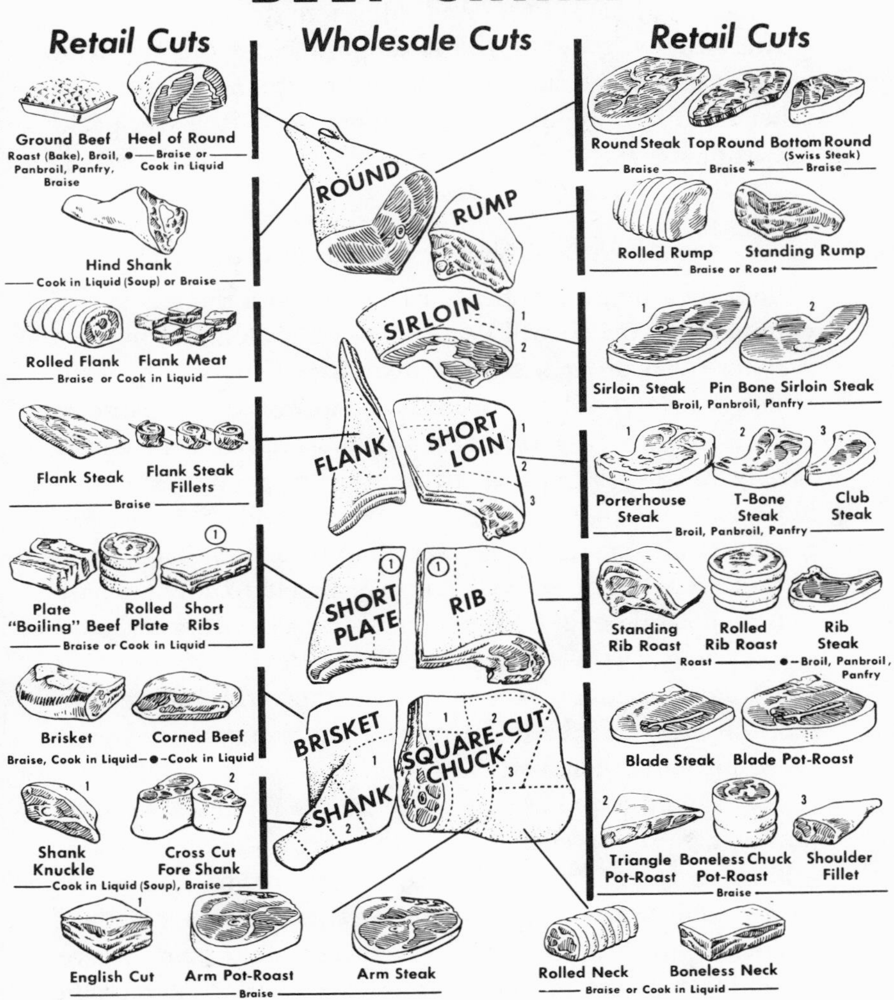

*Prime and choice grades may be broiled, panbroiled or panfried

Courtesy: National Live Stock and Meat Board.

carefully save all the scraps of trimmed-off meat and grind them for recipes such as Mexican chili and Italian spaghetti sauce. For fancier ground beef to be used as chopped steak, I use lean portions of the tender top round.

Incidentally, grinding the meat produces a great deal of heat, so always be sure to chill freshly ground beef or any other meat thoroughly in the refrigerator before you package it for freezing.

How Much Beef Should You Buy? What Cuts?

Your own family's appetite and tastes and the amount of entertaining you do will decide these questions for you.

Before you order a side or hindquarter of beef at what seems to be a very low price per pound "as fallen," pause to consider that a steer is not all sirloin steaks. If your family simply will not eat any of the long list of good inexpensive meat cuts, limiting their beef menu demands to prime rib roasts, broiled porterhouse steaks and an occasional hamburger, you might as well resign yourself to buying only those wholesale cuts from which rib roasts and porterhouse steaks come. The trimmings will supply the hamburger. You won't save as much money, but neither will you receive scornful looks and be told that you are trying to feed your loved ones peculiar food.

Beef can be stored in your home freezer for at least a year, however, so if you cannot resist the bargain implications of a hind- or forequarter, you have plenty of time to conduct a wily educational program calculated to propagandize and influence your family's palate to the sensible point where shank-meat pie or stuffed flank steak will be as gracefully complimented as a rare, juicy T-bone.

The usable meat cuts of the carcass portion you select will vary in weight according to the type or grade of the beef itself. For example, a prime grade steer may weigh in at 1,200 pounds while it is still able to step on a scale. When the carcass is dressed and chilled, only about 547 pounds will be acceptable on your table; this figure includes the kidney, liver and heart, but excludes the bones and fat. In other words, close to 55 per cent of a prime grade steer goes into what is called by-products.

By the way, by-products constitute a major part of the meat industry's profits. *Remember this fact when you arrange for a community locker plant operator or a local butcher to cut up a carcass for you.* He can resell the by-products to a variety of industries which use them in making their marketable merchandise—soap, candles, bone china, violin strings and hundreds of other items. The cut-up man should be made to realize that you are wise to this traffic in *your* animal's valuable waste, and should be persuaded to accept the saleable trimmings in lieu of, or as part payment for, his services.

For the sake of illustration, let's put a side of U.S. Choice or U.S. Good beef into your freezer, assuming that the forequarter and the hindquarter weigh 100 pounds each, dressed.

First, tell the man to remove most of the bones and fat. He will know what you mean, which may not be altogether what *you* think you mean. Some bones and fat are always figured in the total weight. You may trade him all of those he removes, or you may want to retain some of them for yourself. Meaty bones make wonderful stock (see page 130), and beef suet becomes excellent shortening. (See page 149 for recipe and use.)

Boning beef before freezing it makes packaging easier and conserves space within the cabinet. There will always be times when you will decide in favor of bone-in meats, of course, despite the greater care that must be taken in wrapping them and the additional space they require in your freezer. Standing rib roasts of beef, crown roasts of pork and a rack of lamb are well worth the extra trouble and room, especially if these are favorites with your household.

Yield and Uses of the Forequarter

From your 100-pound forequarter of Choice or Good beef, this is approximately what you will get for your freezer:

16 Pounds of Ribs for Roasts and Rib Steaks

Boneless rib steaks: These, comparing very favorably with the more familiar and popular loin steaks, are cut from the rib ends of boneless prime and are most satisfactory if they come from very young beef.

Rib roasts: Have these cut to four or five pounds each if they are to be boned and rolled; otherwise, have the bone-in standing rib roasts cut according to the size of your family's appetite—up to 10 pounds.

7 Pounds of Plate Beef

This may be cut into a bone-in plate section or two, suitable for boiled-beef recipes, or you may have the bone removed and the plate rolled and tied for low-budget pot roasts. The part nearest the rib section may, if you like, be cut into short ribs. Good boneless stew meat can also be cut from this section for you to cube and package in one- or two-pound quantities for freezing.

7 Pounds of Brisket

This somewhat fatty piece may be corned, boiled or ground for hamburger. For corning, which you can do yourself or have done by your locker plant, the bone should be removed. If you grind the brisket for hamburger, do not salt it. After chilling the ground meat in the refrigerator, package it in one- or two-pound quantities, or pat it into individual hamburgers and package in family meal units.

5 Pounds of Short Ribs

30 Pounds of Chuck, or Shoulder

This, the largest wholesale cut of the forequarter, presents many possibilities of dividing into table cuts which should not be overlooked by home freezer owners. From the inside portion of the chuck, which is a continuation of the rib muscle, come blade chuck steaks with a minimum of bone; inside chuck steaks (boneless), and attractive small-sized inside chuck rolls which are very convenient for small families or single meals. The outer portion of the chuck provides small blade steaks and chuck rolls of various sizes. Pot roasts should be cut to about 3½ or 4 pounds each. Excellent Swiss steaks can be made from slices taken from the boneless part of the shoulder.

12 Pounds of Neck

It is for you to decide whether you want all of this cut up for boneless stew or soup or ground for hamburger, or whether your family is fond enough of pot roasts to have it cut into one or two boneless neck rolls which you can have larded with pounded cod fat (from the flank of a hindquarter).

8 Pounds of Foreshanks

High in extractives, this rich meat makes excellent stews, or it may be ground for beef patties. Ask for the bone to use in making stock or soup.

6 Pounds of Lean Meat

Left over from trimming roasts and steaks. Grind this for high-quality chopped steak, remembering to mark it "choice" after you wrap it in one- or two-pound packages for freezing.

9 Pounds of Unavoidable Fat and Bone

Make stock base from the bones, render the fat for suet shortening (page 145).

Yield and Uses of the Hindquarter

If you have never bought beef in quantity before but want to give it a try, this is the quarter which will provide you with the greatest number of the more expensive steaks and roasts which are so popular and desirable. From a 100-pound hindquarter of choice or good beef, this is approximately what you will get for your freezer:

7 Pounds of Hindshanks

Split into two soup bones, trim the meat off and package it in chunks for soup or stringy stew meat. It is exceptionally rich in extractives, and soups made from this meat will jell naturally.

6 Pounds of Heel of Round

This wedge-shaped cut located just above the shank has very little fat and consequently is the least tender part of the round. It makes a very satisfactory pot roast, however, especially if a piece of beef suet is placed in the cavity left when the bone is removed and before the heel is rolled into shape and sewed or tied together. It may be packaged and frozen in one piece, or cut into two smaller pot roasts. If your family is fonder of hamburger than it is of pot roast, this piece may be ground.

26 Pounds of Round of Beef, Top and Bottom

Before you give instructions about cutting up the round for table portions, give thought to your family's preferences. The round, which is divisible into four distinct parts varying in size and tenderness, can provide a quantity of steaks or a quantity of roasts or a combination of the two.

The top round is one large muscle and is the most tender portion. This may be cut into boneless top round steaks or into four- and five-pound roasts.

The bottom round comprises three muscles—the sirloin tip, the eye of round and the bottom round. Of these, the sirloin tip is the most tender and can be cut into very good sirloin tip steaks which, if the carcass is young and prime, may be broiled.

The eye of round, while not so tender, is a very handsome long muscle which, cut across the grain, provides a number of lean eye-round steaks of uniform size just right for a single serving. These can be pan-broiled, braised, with or without pounding, or flattened into extremely delicious Swiss steaks.

The bottom round may also be cut into steaks, but is perhaps most useful when divided into 3½- and four-pound pot roasts.

9 Pounds of Rump

If the bone is left in, have this triangular piece cut into two equal pot roasts; if the bone is removed, have it rolled and tied as a rolled rump roast, cutting it into two equal pieces for packaging and freezing.

14 Pounds of Sirloin

Here, too, your preference for roasts or steaks should guide you in the cutting of this choice portion of a beef hindquarter. Have sirloin roasts cut into four- or five-pound pieces. Sirloin steaks which are cut into at least 1-inch thicknesses will each serve four people. Wrap and package all steaks individually, or, to conserve your packaging materials, pack as many together as your family will use at one meal, being sure to separate each steak with a folded or double thickness of cellophane.

15 Pounds of Short Loin

This portion of the loin is what provides us with T-bone, porterhouse and club (or Delmonico) steaks. Have these cut into the desired thicknesses, package each steak individually and be sure to label the package as to the kind of steak it contains.

7 Pounds of Flank

Have your butcher remove the *flank steak,* which is an oval-shaped boneless piece weighing up to two pounds located between membranes of the flank. Its muscle fibers run lengthwise. It is best to package this piece flat for freezing, and, before cooking it, score it with the edge of a plate or the back of a heavy knife to shorten the muscle fibers. This is the choicest portion of the flank. It can also be rolled around a strip of ½-inch cod fat and skewered one inch apart, slicing between the skewers for flank steak fillets. The remainder of the flank portion, stripped of as much fat and connective tissue as can be gotten off, can be cut into small pieces for boneless stew or ground for hamburger.

4 Pounds of Sirloin Tip

This piece, which is part of the bottom round, can be fashioned into one pot roast or cut into four one-inch steaks.

12 Pounds of Fat, Inedible Bone and Trimmings

Use bone and trimmings for basic soup stocks. Fat may be rendered for shortening.

No Filet Mignon?

It would be my considered advice not to have your hindquarter cut in such a way as to yield this piece, choice though it is. It is a comparatively small portion of the loin which runs through the sirloin, porterhouse and T-bone steaks and it seems a shame to sacrifice the best part of them to obtain a fillet which is at most six or seven pounds, ordinarily. It is because of its central location in the loin and the havoc it wreaks on surrounding cuts that it costs a king's ransom in fancy markets and restaurants. If you want to treat your family to a luxurious filet mignon dinner, it might be better to ask your butcher or locker man to pick up a fillet from his wholesaler for you.

How to Make Your Own Suet Shortening

Suet is the hard fat found around the kidney and loins of a beef animal, corresponding to the leaf lard of hogs. Do not use the softer, coarser-textured exterior fat for the following recipe:

Chop or grind suet while it is very fresh and then put it away, lightly covered, in your refrigerator. Let it stay under refrigeration for at least twenty-four hours. Try it out slowly over low heat. Stir the melted fat frequently and let it cook, still slowly, until it is clear and smooth and the bits of unmelted fiber are crisp and brown. Strain the liquid through several layers of cheesecloth, then measure it.

Add to the melted suet half of its measure of ordinary vegetable salad oil. Chill the mixture quickly, stirring it occasionally as it hardens. Keep the shortening in the refrigerator or, packaged in small quantities, in the freezer.

Two pounds of chopped suet, rendered, plus half of its melted equivalent of vegetable oil will generally yield between two and three pint containers of suet shortening.

This shortening is creamy and easy to work with and is especially good for general baking. Because it is considerably richer than the commercially manufactured solid shortenings, a scanter measure should be used than the amount called for in recipes.

COOKING FROZEN MEAT

Recommended cooking methods for the various grades and cuts of frozen beef and other meats are given in the charts commencing on page 334.

8. Ventures in Veal

Like the beef in the previous chapter, veal is subject to the same federal inspection standards and grades which are applied to cattle carcasses sold for public consumption. The three characteristics of conformation, finish and quality described on pages 129 and 130 apply also to this meat.

Although most of the meat in this class is usually advertised and sold as *veal,* the government recognizes two other classifications—*calf* and *baby beef.*

Veal is the flesh of very young animals, usually between six and fourteen weeks old. In the better grades, these youngsters are fed principally on milk or milk substitutes. The meat is extremely fine-textured and has a characteristic pinkish-gray color. There is very little exterior fat.

Calf is from young animals which have grown somewhat beyond the veal age but have not yet arrived at the maturity of beef. The flesh of calf is firmer and coarser than veal and deeper red in color. Fat is slightly more developed on the outer and interior surfaces.

Baby beef, from older calves, comes from well-bred, specially fed animals from seven to fifteen months old. These carcasses are judged and graded according to the beef standards of conformation, finish and quality. The meat, however, is likely to be more delicate in flavor than beef because of the animal's youth.

Government grading standards for veal are Prime, Choice, Good, Medium (or Commercial), Utility (or Common) and Cull, the last named being equivalent to the Cutter and Canner grades of beef and used almost exclusively by the meat canning and meat specialty product industries.

Very few veal or beef carcasses that reach the market qualify for the grading stamp which identifies it as Prime. The highest grade you will find in wholesale meat warehouses, community locker plants or local butcher shops from which you purchase meat in quantity is *U.S. Choice Veal,* which is recognizable by the following characteristics:

Conformation—blocky, compact, with well-fleshed legs, ribs and loins.

Finish—because of its immaturity, this applies not to exterior fat (as it does in relation to beef) but to the amount, color and type of meat and how it is distributed over the frame. U.S. Choice veal has firm, brittle, off-white fat in ample quantity around the kidney and pelvic region.

Quality—preponderantly lean meat, light pinkish-gray in color and exceptionally firm and velvety in texture. The meager amount of exterior fat which is characteristic of Choice veal is clear, firm and almost pure white. There is no marbling, as in beef, and the bones are porous and very definitely red in color. The knuckle and joint ends of veal bones are soft and pliable.

In all grades of veal, especially the choice or younger carcasses, there is a relatively high proportion of bone to muscle, which means that the wholesale (or primal) cuts available to you in bulk purchases are more limited than they are for beef.

Where and How to Buy Veal in Quantity for Freezing

As with beef, the good sources are retail or wholesale butchers and community locker plants, also neighboring cattle ranches and farms with modern facilities for slaughtering and chilling.

If your family is especially fond of veal, a whole carcass is not too much to buy and freeze for a year's supply, for the entire young animal, when dressed, will weigh little more than 100 pounds, and sometimes less. From 50 to 60 per cent of the liveweight goes into by-products of the meat industry. You may, of course, buy a side, a hindsaddle or foresaddle (sometimes called the "rack") or one of the wholesale cuts as shown in the chart on page 149. Generally speaking, however, the more you buy of this meat at one time, the more you will save in dollars and cents. Held at low temperatures, veal keeps especially well in home freezers and offers many opportunities for varied and delicious menus.

If you decide to buy a whole carcass from its original source, a ranch or farm, insist on receiving the liver and sweetbreads (thymus gland). These are ordinarily omitted when you buy a dressed carcass, but they make wonderfully savory and nutritious recipes. See the discussion of Variety Meats beginning on page 166 for information about prepara-

VEAL CHART

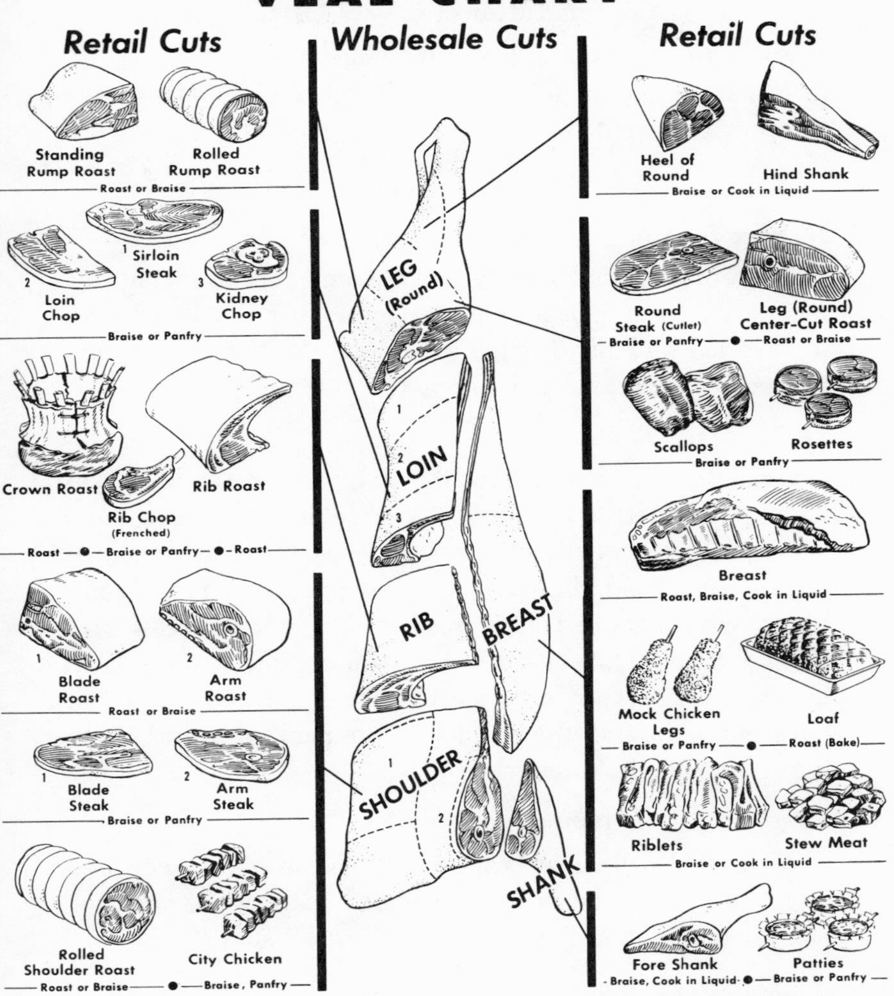

Courtesy: National Live Stock and Meat Board.

tion and recommended length of storage time in your home freezer.

Yield and Uses of a Veal Carcass

From the average carcass of choice veal, weighing approximately 100 pounds, you will get the following cuts for your freezer:

Hindsaddle

15 Pounds of Leg Roasts and Cutlets

Have the legs cut into two five-pound roasts, slicing five pounds off the rounds for cutlets. Cutlet veal may be broiled, or breaded and pan-broiled, or it may be sliced very thin for recipes such as veal scallopini.

11 Pounds of Shanks

No matter where you have this cut, some of the bone will inevitably be left in for you to use profitably in making soup or basic stock (see directions on page 126). Trim the bones closely and grind the meat for veal patties or to be combined with chopped beef and pork for croquettes and sausage meat. Do not salt ground meat, and always keep it in the refrigerator for at least twenty-four hours after you have put it through the grinder before packaging and freezing it in meal-size amounts.

12½ Pounds of Rumps

Since you will have two, you might want to keep one intact and have the other one boned and rolled, both for Sunday-dinner roasting.

6½ Pounds of Loin

This choice section can be retained intact as a sirloin roast, or it can be cut into steaks or chops for faster cooking. This portion of a veal carcass contains the kidney, which you may, if you like, have removed for separate packaging and freezing. Loin veal chops which contain slices of the kidney are attractive and exceptionally tasty, and you may prefer to have these chops cut in this manner.

Foresaddle

11 Pounds of Rib (or Rack)

When this section is divided into two roasts of equal size, they compare to the standing rib roasts of prime beef. One of the half-racks may be prepared as a handsome crown roast, or you may decide to cut one or both into delicious veal rib chops. If it is your decision to retain the rack for roasts, ask that the bones be kept fairly long so that they may be "frenched" to form a crown. When specifying chops, have the bones cut short and use the leftover pieces to add to soups or stews or to simmer for the basic stock you freeze.

13 Pounds of Shoulder

This fleshy portion of veal may be cut to provide a blade roast, an arm roast and a rolled shoulder roast, or it may be divided among two roasts and several arm and blade steaks, as you prefer.

9½ Pounds of Breast (two pieces)

Stuffed breast of veal is so good that you may want to freeze both pieces for this purpose. If you do, be sure to cut the pockets before packaging and freezing. If you prefer, save just one side of the breast in a single piece for roasting, and have the other piece cut into riblets or into stewing cubes.

8½ Pounds of Neck (and flanks, from the hindsaddle)

This lean meat, removed from the bones, may be cubed for stewing or for recipes like goulash or it may be ground for veal loaf, patties and mock chicken legs. Wrap in one or two-pound packages for freezing.

1 Pound of Kidney (if not cut in with loin chops)

Trim fat away, chill the kidney thoroughly before freezing, and store for less than three months (see page 125).

9. A Larder Full of Lamb

Most people, when they think about lamb, think in terms of roast lamb or lamb chops. This tender, young, flavorful creature produces more varied cuts and main-course possibilities than is generally realized, however, and most of the cuts are suitable for roasting, broiling and short braising, while others are excellent for long braising, stewing or cooking *en casserole.*

Mutton is not especially popular in this country. Only 10 per cent of all sheep raised for meat are permitted to mature to the mutton stage, 90 per cent being slaughtered as they reach the classifications of Hothouse lamb, Genuine Spring lamb, Spring lamb and Yearling lamb.

Hothouse and Genuine Spring lambs are slaughtered before they have reached the weaning period, and consequently are sometimes referred to as "milk lambs."

Spring lambs are slaughtered after weaning up to the age of twelve months.

Yearling lambs are more than a year old, but usually less than two.

The United States Government grades for lamb and mutton, based on the carcass' conformation, finish and quality as for beef and veal, are Prime, Choice, Good, Medium, Plain and Cull. The three top grades represent animals that are compact and blocky with short, plump legs, heavily fleshed loins and flanks.

Freezer owners will find it economical to purchase lamb by the carcass or in the larger wholesale cuts, which are the legs, loins, hotel rack or the foresaddle with the rack removed. You will do well to be guided by the advice of your honest butcher when you buy lamb, for it is not easy to recognize the quality of this meat.

If you have always been in the habit of purchasing lamb from fine butcher shops, you may never have seen the tough, papery covering over the rack and loin which is called the "fell." Most meat markets remove this membrane for the sake of appearance and attractive display, but distinguished chefs know that the fell actually improves the flavor of cooked lamb by retaining the meat's juices. It also acts as an insulating agent and shortens cooking time and helps preserve the shape of the cut during the roasting process.

If you purchase a whole carcass for freezing be sure to ask for the liver, heart and kidneys and freeze these vitamin-rich variety meats separately for short-term storage, cooking them later in a number of delicious main-course dishes. They are especially good in stews or combined with spicy sauces *en casserole.*

Yield and Uses of a Lamb Carcass (40 Pounds)

12½ Pounds of Legs

If you decide to keep these whole for roasting, you might like to have one "frenched," a style which exposes the bone

LAMB CHART

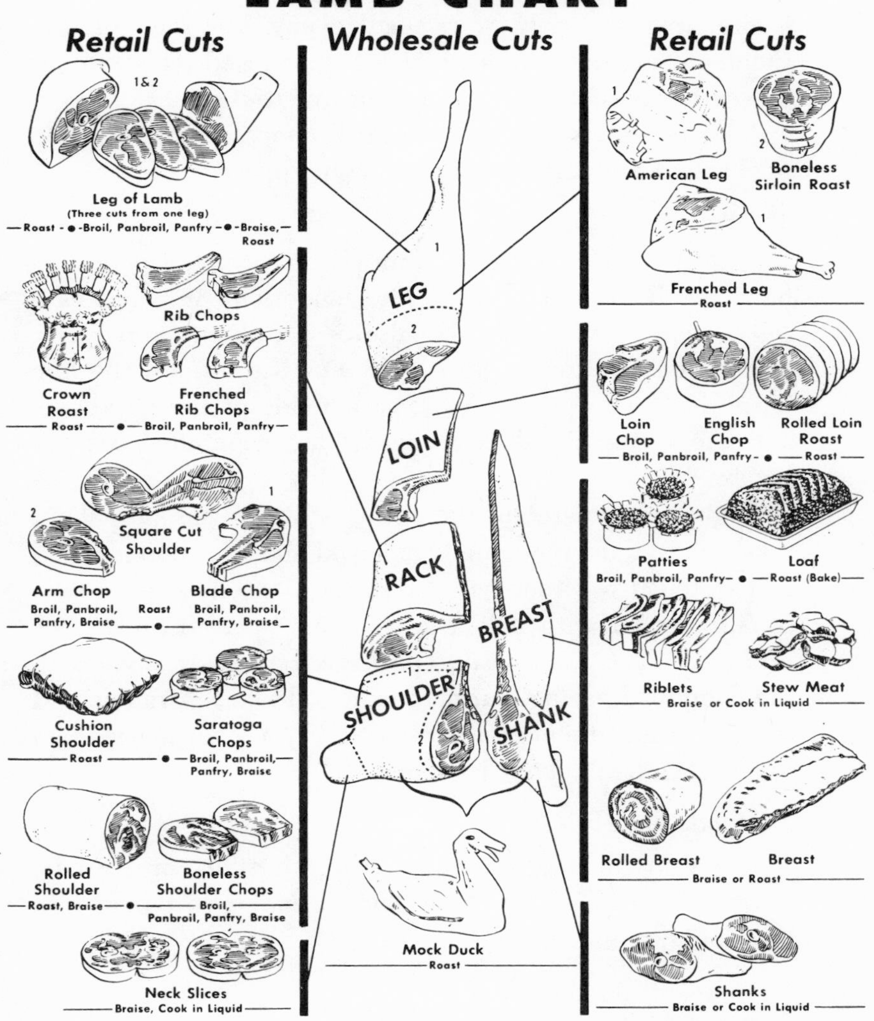

Courtesy: National Live Stock and Meat Board.

at the lower end of the shank and provides a convenient handle to be grasped while carving.

The other leg, prepared according to the American style, is preferred by cooks with small ovens. For this cut, the shank meat is removed at the joint and tucked into a pocket on the inside of the leg to prevent overcooking.

A leg of lamb may also be cut into chops, which resemble beef round and large sirloin steaks in appearance.

4½ Pounds of Loin

This may be retained as an unsplit loin roast, or you may have the bone removed and the tenderloin prepared as a rolled roast. This section may also be cut up into loin chops, either American or English style. For the former, the loin is split into two pieces, one on each side of the backbone, and the chops sliced in wedges which contain T-shaped bones.

English chops are cut the entire width of the loin, about two inches thick, the bone is removed and the chop is rolled and fastened together with a skewer. This style of chop is usually served with half a lamb kidney placed on top for broiling, but it is not wise to package the kidney along with the chop for freezing. Package lamb kidneys separately, and be sure to slice them before freezing if you plan to use them for English lamb chops later on.

4 Pounds of Ribs (Rack)

This may be roasted intact, English style, or may be fashioned into a crown roast. The rib may also be cut into frenched or plain chops.

9 Pounds of Shoulder

This may be cut in a variety of ways, depending on your preference and your mood for new adventures in cooking. If your meat source employs a skilled butcher, you might ask him to make a mock duck for you, a task which amounts to meat sculpture for the completed roast looks for all the world like a duck, with cranberry halves for eyes and a piece of cartilage forming the tail.

The most popular roasts from this portion of the lamb are the square-cut shoulder, the cushion shoulder and rolled shoulder. This cut can provide, instead of roasts, a number of attractive boneless shoulder chops, arm or blade chops and the picturesque, skewered Saratoga chops made from the tender inside shoulder muscle.

4 Pounds of Breast

This narrow strip of somewhat coarse but tasty meat may be roasted with or without a pocket for stuffing, or it may be boned and rolled. The four pounds of breast meat may also be cut as riblets for barbecuing.

3 Pounds of Stewing Meat from Neck and Shanks

Because lamb stew and shepherd's pie are popular with many families, you may want neck and shank meat cut into cubes for these dishes. However, the neck may also be cut into single or double slices for braising or casserole courses, and the shanks may be retained for making mock chicken drumsticks.

1½ Pounds of Variety Meats

Liver, kidney and heart.

1½ Pounds of Fat, Bone and Trim

Save as many bones as you can from a lamb carcass and simmer them down for stock to be used in many of the soups for which lamb stock is desirable.

How to Make Lamb Soup Stock for Freezing

Cut meat from lamb bones and brown well in a little fat. To a pot containing six pounds of lamb bones add the browned meat, six quarts of cold water, ⅔ cup of chopped celery, ⅔ cup of diced carrots, ⅔ cup of chopped onion, ½ cup of chopped parsley, 10 whole cloves and two bay leaves. Cover the pot and simmer the broth for about four or five hours. Strain the liquid and chill it. Remove fat and strain again.

Freeze in ice-cube trays or small containers, removing the blocks after they are frozen and packaging them in polyethylene bags. Add salt to taste when making soup rather than before freezing.

When cooking lamb according to the methods recommended on page 335, omit sage from stuffings. Mint sauce, of course, is traditional with this meat; try seasoning it with marjoram for an interesting new taste.

10. Going Whole Hog with Pork and Ham

Table pork in its many forms and delicious cuts comes from five different classes of hogs raised for meat—barrows, boars, gilts, sows and stags. Barrows are castrated young males; boars are mature male breeders; gilts are young females before breeding; sows are mature female breeders, and stags are older males which have been castrated after breeding.

Most of the pork and ham which is offered to us in retail or wholesale meat markets is from barrows, gilts and occasionally sows; stag and boar meat is usually sold to packers who manufacture sausage and luncheon meats.

The government grading of U.S. inspected pork is somewhat different from the standards for beef, veal and lamb, for the wholesale cuts are judged on their individual desirability rather than on the conformation, finish and quality of the entire carcass.

If you do go whole hog in your purchases and wish to freeze such things as smoked hams and bacons as well as fresh pork roasts and chops, you will do better to rely on your community locker plant operator or a local curing center to provide their services, unless you happen to have an old-fashioned smokehouse in your back yard.

Before you purchase a whole carcass, it is wise to find out, if you possibly can, how the animal offered to you was fed during its lazy lifetime. No other meat depends so much for its flavor, texture and quality on the type of food which was consumed by the beast on the hoof. This is principally because beef, veal and lamb are the products of pasture-grazing animals which feed on natural grains and on milk and milk products, whereas pigs are pigs and can be kept as happy with a diet of garbage, scraps and peanut shucks as they are with corn and mash.

Needless to say, it is the corn-fed or protein-enriched mash-fed hog which has the firmest, finest-grained flesh of a grayish pink color marbled with flecks of fat and with a uniform covering of firm, white fat on the exterior surfaces. These are by far the best eating hogs. Animals which feed on peanuts or other oily food are inclined to produce somewhat oily meat, frequently tasting of peanuts.

Rooting pigs, left to their own gastronomic devices, would probably show their native good sense by foraging for edible roots, grains and berries after they were rejected by their nursing mothers. But if the poor things are enclosed in a grassless pen and thrown nothing but table scraps and kitchen leavings, they will soon learn to eat unimaginable things and will inevitably grow up with more fat than edible muscles on their frames.

A good dressed hog will be a well-fed animal slaughtered

PORK CHART

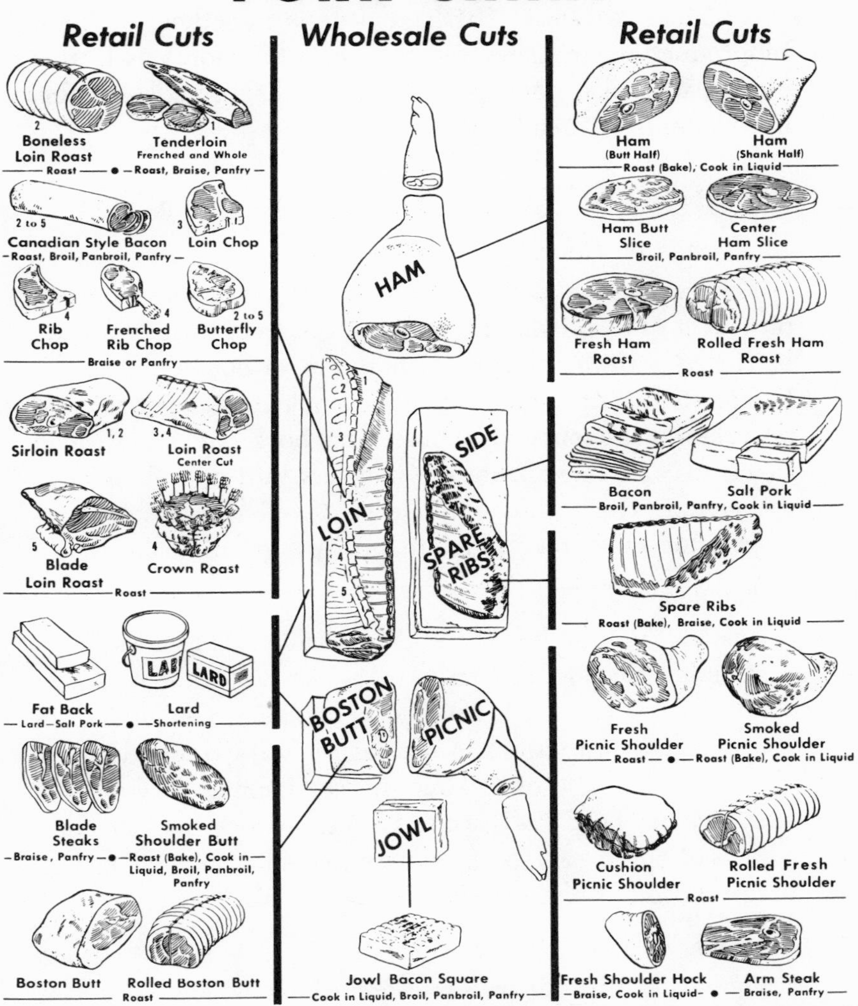

Courtesy: National Live Stock and Meat Board.

at the age of six to seven months and weighing about 225 pounds on the hoof. Dressed, which means that it has been scalded, shaved and scraped, the carcass will weigh about 175 pounds, the rest of it having gone into those valuable by-products which later reach us in retail stores as leather gloves, footballs, photographic film and hundreds of other useful items of merchandise.

Yield and Uses of a Small Pork Carcass (125 Pounds)

25 Pounds of Hams

The two hams, weighing about 12½ pounds each, can provide a splendid variety of meals. Roasts of fresh ham may be frozen with the fairly large bones in, or boned and rolled. Either or both of the hams may be smoked whole or divided into the component butt, center and hock cuts. Fresh ham steaks may be taken from the boneless butt as well as from the center. Hock ends may be split and the knuckles removed for small roasts and to provide grinding meat.

20 Pounds of Bellies

This section of a hog is frequently called "side meat," and it includes the fat-and-lean strips which we know as bacon after it has been cured, usually by smoking. Depending on your family's taste and the recipes for salt-cured side meat you are in the habit of making, this section may be frozen in meal-size packages as fresh *sliced* side pork, pickled side pork, salt-cured side pork or as cured and smoked bacon. Freeze bacon in uniform slices for easier use later.

3 Pounds of Spareribs

These constitute the ribs and breastbone which have been removed from the bacon strip. Have them split into the size you are accustomed to cooking, and freeze them fresh in two packages of equal size for baking or barbecuing.

20 Pounds of Loin

This versatile, popular and delicious cut can, according to the way it is divided, provide you with no less than a dozen kinds of main courses. The ham end may be cut off, boned and rolled into a very satisfactory sirloin pork roast; or the lean portions, also boned, may be sliced and pounded to make a number of pork tenderettes. The shoulder end of the loin may also be converted into attractive boneless roasts or into country-style backbones.

The tender and succulent center cut of the loin makes an exceptionally choice pork roast, whether it is permitted to remain as a rack or is fashioned into a regal crown roast. Instead of freezing the loin as roasts, you may prefer to have it cut into frenched chops, butterfly chops or just plain pork chops.

For delectable variety, you may ask the butcher to cut the tenderloin portion out separately. Cut this section across the grain into two-inch pieces, place each piece between two pieces of laminated freezer paper and flatten it to about ½-inch thickness with the side of a cleaver, a mallet or any broad, fairly heavy instrument.

By cutting the tenderloin into four-inch pieces and splitting each section lengthwise without cutting them completely in half and then flattening them as above, you have frenched tenderloin strips with which to make pork birds.

24 Pounds of Shoulder

The arm and shank section of one shoulder may be prepared as a fresh (or smoked) picnic shoulder and the other as either a boneless roll or a cushion roast into which has been cut a pocket for stuffing. These will be fairly large, however, and you may prefer to ask your butcher to cut the shoulders into several three- or four-pound roasts, with some of the meat cut into arm and blade pork steaks.

12½ Pounds of Fat Back

The lean portions of this section, along with other trimmings, may be ground for sausage and packaged, spiced if you like but with salt omitted, either as meat patties or stuffed into casings. The fatty part can be cured as salt pork for flavoring, called for in many recipes, or may be smoked for a rather heavy type of bacon.

2½ Pounds of Jowls

Trimmed square, cured and smoked, this is frequently served as heavy bacon (better cut it to uniform slices before freezing). Or, allowed to remain unsmoked but cured with dry salt, it can be used in recipes calling for salt pork.

3½ Pounds of Knuckles and Feet

It is best to pre-cook and pickle these.

4 Pounds of Leaf Lard

The fat found on the inside of the body cavity makes the finest and purest lard imaginable, and therefore care should be taken either at the butchering place or at home to keep it separate from the carcass' other fat. An economical and extremely easy-to-use shortening, lard has high nutritive value and imparts delicate flavor to other foods when used as a frying fat. Rich in energy-producing nutrients and 97 per cent digestible, lard is preferred by many home economists and highly rated cooks for all baking and most cooking which calls for additional fat.

8 Pounds of Trimmings and Extra Fat

Salvage what you can of the trimmings for sausage, or make your own Philadelphia scrapple. Incidentally, another version of scrapple has for its main ingredient a pig's head.

A second-quality lard can be made from the extra fat, but should not be kept in the freezer for more than three months.

How to Make Your Own Lard

Render lard cut into small cubes slowly over low heat until cracklings rise to the top crisp and brown. Strain and chill it thoroughly before packaging it for freezing. Never pour hot lard into wax-coated containers. I have kept well-wrapped lard in my freezer for more than six months.

Four pounds of leaf lard, when rendered, will yield approximately three pint containers.

Pork Facts

Food-chemistry research laboratories have discovered that a very definite value occurs when pork or pork products are frozen in the fresh state. When this meat is held at 0°F. for three weeks or more, any of the trichinosis organisms which occasionally infest pork and may be present, are killed.

Unless you plan to consume large quantities of pork within a comparatively short period of time, it is best not to smoke or grind too much of the carcass, for smoked and ground meats, especially pork, do not preserve as well in the freezer as do their fresh or whole counterparts.

11. Variety Meats

All of the meat-providing animals we have discussed in the previous chapters include, as standard equipment, various organs and parts which are not classified as regular meat cuts by butchers or the meat industry. They are known as "variety meats." These—the brains, hearts, kidneys, livers, sweetbreads, tongues, tripe and oxtails—are not only good eating when imaginatively prepared as delicacies, they are also among the most highly nutritive foods in our larder.

Any modern discussion on nutrition you may read, either in the popular press or in the literature of the medical and scientific professions, will most usually comment, when describing certain of the known vitamins and minerals present in food, "important for liver function, heart action, brain tissue, etc." Nature, in utilizing essential food elements, pays particular attention to the healthy operation of the body's vital organs and, a wise and provident manager, stores up quantities of enriching vitamins and minerals within these organs.

Primitive peoples, innocent of civilized man's ruthlessly refined and devitalized foods, believe that eating an animal's organs will magically transfer to themselves the beast's desirable characteristic qualities. The heart of a lion, they calculate, will give them fearlessness and strength; the brain of a fox, cunning.

Modern medical science, while not subscribing to such magical powers, nevertheless attributes to the organ meats great preventive and curative abilities. Indeed, many of the physician-prescribed vitamin capsules we buy from drug stores because we have consistently refused to eat the variety meats are synthesized from the very animal organs we have disdained in our self-indulgent diets, for we have grown to delude ourselves that food exists to satisfy appetite. Babies, animals and primitive tribes possess instinctual knowledge of the functions of food: They know that it satisfies hunger.

Among your opportunities for better nutrition and consequently better health, as a home freezer owner, is the quality of the meals made possible by your "domestic supermarket." If your freezer and your menus include carefully husbanded and appealingly prepared servings of organ meats, your family will benefit in health and you will soon earn the deserved reputation of being a talented specialty cook.

Especially if your quantity meat purchases are made in carcass form, or if you live near a source of freshly slaughtered meat, you can arrange to buy freezable quantities of the variety meats for meals which are not only highly nutritious and savory, but also extremely economical.

Our peculiar purchasing standards as a nation, set by a public which demands porterhouse steaks and scorns the frequently more nutritionally valuable chuck or shank portions have, for example, forced the immature and relatively less enriched calves' liver to a premium price it does not intrinsically warrant; beef, pork and lamb livers, equally edible and much more nutritious, cost far less.

Freezing Variety Meats

Brains (beef, veal, pork, lamb)

Nutritive value: Rich in the B vitamins niacin and thiamine.

These are tender, soft in texture and of a very delicate flavor. For home freezing, avoid any which contain blood clots. Package separately and label clearly according to classification (whether beef, veal, pork or lamb).

Because most recipes call for pre-cooking, you may prefer to freeze brains which you have prepared in advance for later recipes. To pre-cook, wash and simmer brains for twenty minutes in water to which you have added one tablespoon of vinegar and one teaspoon of salt for each quart of water in the cooking vessel. Drain, cool, refrigerate, and package in meal-sized portions in laminated freezer bags or in aluminum foil.

Hearts (beef, veal, pork, lamb)

Nutritive value: Very rich in B vitamins riboflavin and niacin and in iron; good source of thiamine.

Remove large arteries and trim off any hard parts, then wash thoroughly.

A beef heart will serve up to twelve persons, pork and veal hearts will serve two, and a lamb heart one. Leave the hearts whole or cut into pieces according to the recipes for which they will later be used. Package in laminated freezer paper or bags, or in aluminum foil.

Kidneys (beef, veal, pork, lamb)

Nutritive value: Exceptionally rich in riboflavin, niacin and iron; good source of phosphorus.

Kidneys are highly prized by epicures and famous chefs for their distinctive flavor. Remove all kidney fat and tubes, wash and split through the center. Wrap individually. Beef kidneys may be cut into meal-size portions or into pieces for stews and pies.

Livers (beef, veal, pork, lamb)

Nutritive value: Exceptionally rich in riboflavin, niacin, Vitamin A and iron.

Wipe with a damp cloth. It is not necessary to scald liver. Slice beef and veal livers about ½-inch thick and slip-sheet with folded cellophane or glassine paper before packaging in meal-size portions. Pork and lamb livers may be sliced or cut into pieces, depending on their intended use later.

If you wish to make liver loaves or patties, drop the liver into hot water and let it simmer gently for a few minutes to facilitate grinding. Chill or refrigerate thoroughly before packaging and freezing.

Sweetbreads (veal, pork, lamb)

Nutritive value: Good source of Vitamin C and niacin.

Many people mistakenly think that this is the pancreas. Actually, it is the thymus gland, situated, in young animals only, in a place which corresponds to a human being's thorax. Its function is concerned with growth and blood formation and, in nature's marvelous way, disappears in the adult animal when its services are no longer required. (This, incidentally, affords a good detective clue to those of you who examine an undressed carcass to determine its age. If the thymus gland is present, it is a young animal.)

A great delicacy among discriminating gourmets, sweetbreads are similar in texture to brains, and are equally tender and delicately flavored. If planned recipes call for pre-cooking—and most do—you may wish to pre-cook sweetbreads before packaging and freezing. Follow the directions given for pre-cooking brains. If any membranes are present, they may be removed after pre-cooking. Chill or refrigerate the drained sweetbreads and package them in meal-size portions in laminated freezer paper or bags, or in aluminum foil.

Tongues (beef, veal, pork, lamb)

Nutritive value: Rich in iron; fair source of niacin.

Beef and veal tongues, which form the base of many delicious and hearty recipes, may be frozen fresh, although the preferred method is to have them corned or smoked in advance. Pork and lamb tongues, of smaller size, may be pickled in brine and either frozen or put into vacuum-tight glass jars. Freeze beef and veal tongues whole, including the roots, wrapping them securely in laminated freezer paper or in aluminum foil.

Tripe (beef only)

Honeycomb tripe is the muscular lining of a beef animal's second stomach; smooth tripe, which is pocket-shaped, is the stomach's wall. Either of these may be pickled or canned, rather than frozen. However, freezing is an excellent way to preserve tripe which has been pre-cooked. Cover tripe with water and simmer for about two hours. When finished, the surface will have a clear, jelly-like appearance. Drain and refrigerate. Cut into bite-size pieces. Package in rigid containers or in laminated freezer bags.

Oxtails (beef only)

Not properly an "organ meat," of course, but the very tasty base for hearty soups and stews. Look for a generous covering of white fat around the jointed bones. Cut apart at the joints and package in laminated bags.

Special Note

These are the most delicate and consequently most perishable of all meats. Do not, therefore, plan to freeze and store at one time more than your family will consume during the recommended storage periods. See the chart on page 125 for each variety. Be sure to select for packaging only the sturdiest materials which are moisture-vapor-proof and contain effective oxygen barriers.

12. Poultry in Your Freezer

"Someday, I'm going to have a little chicken farm" is a composite American dream that can come true for home freezer owners, even for those who have back yards not much bigger than a handkerchief. Several freezer families of my acquaintance, as a matter of fact, have successfully raised chickens for eggs and meat in specially rigged-up basement coops.

Scientific information on the breeding, housing, feed, care, killing and dressing of chickens can be obtained from the United States Department of Agriculture as well as from the Department of Poultry Husbandry of most state colleges of agriculture.

A good many breeders, not all of them professional, are going in for experimental raising of new types of poultry such as all-white-meat chickens and capons, and lightweight turkeys of from five to ten pounds, all of them excellent for freezing at home.

Chicken-raising can provide tremendous satisfaction for those who are interested in the fascinating subject of nutrition and its effect on the living body. By controlling and varying the feed, one farmer succeeded in breeding hens who produced eggs with blue yolks! Your purpose may not be so exotic, and neither is mine . . . but someday, I'm going to have a little chicken farm. . . .

More seasonal than four-legged food, a great deal of poultry is consumed in the localities where it is raised. Your local department of agriculture or county agent will be able to tell you the best seasons for purchasing chickens and other poultry in quantity from farms and vendors in your community. Do not hesitate to ask for this information, for considerable savings can be effected if you buy chickens by the dozen and holiday turkeys in pairs at certain times during the year. By finding out when a turkey farmer, located a pleasant morning's drive away, planned to cull his flock, I was able, one year, to serve an enormous Thanksgiving turkey which cost me less than 40 cents a pound, while less opportunistic friends and neighbors were complaining bitterly that their birds had cost them twice as much.

Buying large birds—turkeys and chickens—and cutting

them in half for family meals will bring down the per pound price of your poultry.

I have had the pleasure, too, of meeting the prospective mother of my freezerful of large roasters before she was ever introduced to the rooster who sired them. Both parents were fed an impressively nutritious diet and the hatched chicks led a sedentary and pleasurable life of gourmet quality. Their meaty breasts, I swear, were interrupted only by their plump and saucy tails.

Another time, I put my poultry requirements into the keeping of a friendly local retailer, who has a surgeon's eye and an obstetrician's instinct. He must, for when I brought the hens home to clean and freeze, I found a bonus of at least a dozen egg yolks included in my purchase.

Chickens for the Freezer

Chickens for freezing may be purchased in quantity either alive, New York dressed or eviscerated (full drawn), and are classified in the following manner:

Squab broilers: From 6 to 8 weeks old, weighing up to 1½ pounds.

Broilers: Young chickens (8 to 12 weeks old) of either sex which have soft flesh and tender skin. The breast bone is flexible. Broilers weigh around 2½ pounds each.

Fryers: Young (from 12 to 20 weeks old) and with the same physical characteristics as broilers, but weighing up to 4 pounds maximum.

Roasters: Both roosters and hens from 5 to 9 months old. Soft flesh, tender skin, with slightly more rigid breast bone than broilers or fryers. Roasters usually weigh from 3½ pounds up.

Capons: Young (7 to 10 months old) males which have been unsexed. These have very little comb development, and usually weigh more than 4 pounds each.

Stags: Young, virile males with fairly well developed combs and spurs. The flesh is somewhat toughened, and the breast bones are rigid.

Fowl: Mature females with hardened breast bones.

Cocks: Old males. The flesh is unmistakably toughened, and breast bones are hard.

LIVE CHICKENS

Buying your chickens for the freezer while they are still alive is one way of being sure to get juicy, meaty birds for your family, for you can provide them with the proper nutrition to insure their tenderness and delicacy. Confine the flock to a small enclosure for two or three weeks, feeding them on a milk diet. Withhold all feed for 18 to 24 hours before killing the birds in order to empty their crops.

After killing and bleeding a chicken according to the methods recommended by the U.S. Department of Agriculture, remove feathers either by dry-picking or by holding the bird in water which has been heated to 120° to 130°F. for about one minute, or until the wing and tail feathers can be pulled out easily. Pluck cleanly, removing pin feathers by singeing. If you singe chickens over an alcohol flame (3 tablespoonfuls of rubbing alcohol in a small tin can) you will find that you can prevent the smoking and discoloration which sometimes result from gas-flame singeing.

Remove body heat by submerging the chickens in ice water until flesh is firm. Draw the chickens, removing and discarding the head, feet and oil sac. Or, if you prefer, retain the feet for use as a base for jellied soup stocks. Take out the gullet, crop and windpipe. Dislocate the lungs and heart, then remove the entrails through an incision around the vent. Take care not to break the intestines while they are still in contact with the body of the chicken.

Separate the heart, liver and gizzard from the entrails and

wash them thoroughly. The gall bladder, which is attached to the liver, must be removed very carefully in order not to break it and spill its bitter fluid over the liver.

Wash and drain the bird thoroughly, and chill it in the refrigerator for several hours before packaging and freezing it.

NEW YORK DRESSED CHICKENS

In most communities of the country, these are birds which have been bled and plucked, but not drawn. Their per pound price is, accordingly, lower than fully prepared chickens. Draw according to the foregoing instructions, wash, drain and chill in the refrigerator.

EVISCERATED CHICKENS

These are full-drawn and presumed to be ready for seasoning and cooking, although I rarely cook or freeze one without subjecting it to personal critical inspection for absolute cleanliness of the cavity.

Preparing Chickens for Freezing

BROILERS

Split in half lengthwise, or cut into quarters. Separate all pieces with a folded or double sheet of cellophane to speed later thawing and facilitate handling for cooking. Wrap giblets separately in cellophane or package them in polyethylene or laminated bags, freezing them separately for the shorter-term storage they demand. Wrap all pieces of one broiler together in cellophane, laminated freezer paper or aluminum foil. If you use cellophane or foil, protect the package with an outer-wrap of heavy kraft paper or stockinette.

Broilers may also be packaged in polyethylene or laminated freezer bags of good quality. If in polyethylene bags, force all air out and twist the top of the bag into a goose neck before fastening it with a rubber band.

Date and label the package and freeze it immediately at 0°F. or below. After the individual broilers have been solidly frozen, several of them may be gathered together and put in a large polyethylene bag for convenient handling in storage.

FRYERS

Disjoint according to your family's preference in pieces. Legs may be left whole or divided into drumsticks and thighs. Back and breast may be left whole or divided into two or four parts. Separate the pieces with double thicknesses of cellophane, wrap giblets separately, and package either as one cut-up complete fryer or divide several into parts—three or four breasts in one package, a half-dozen drumsticks in another, etc.

Use laminated freezer paper, cellophane, foil, bags or lined trunk-opening boxes for parts. When you label these packages, be sure to identify the pieces in each one by name, i.e., "four drumsticks, four thighs," "two whole breasts," etc.

ROASTERS AND CAPONS

Wrap giblets separately and replace them in the cavity. Truss legs and wings to the body for easier wrapping. Package in laminated freezer paper, cellophane or foil protected by stockinette, or in polyethylene bags, excluding all air. Pad bony and protruding parts with small wads of cellophane or paper before wrapping to prevent package punctures.

While roasters may be stuffed before freezing, I do not recommend it except for short-term storage of no longer than three or four weeks, unless care is taken to avoid heavy spices in the stuffing.

FOWL AND COCKS

Cut for fricassee and freeze parts together without separating them with cellophane. Or, stew the birds whole until they are tender, retaining the liquid in which they have been cooked as the basis for soup or basting stock. Cut the meat from the bones in cubes or slices for salads, creamed dishes and sandwiches. Pack the cooked chicken meat in laminated freezer bags, lined boxes or in plastic containers.

Chicken Hints

When I freeze a batch of chickens, I usually package half of them whole, for roasting, the rest cut up into parts for broiling or frying.

Chicken livers are a great favorite in my house and buying chickens a dozen or more at a time, as I do, provides a bonus meal or two every time the supermarket has a sale of broilers or fryers. The livers are frozen separately.

Additional bonuses are stock made with chicken necks, and giblet gravy from gizzards and hearts. The stock is usually made the day I bring the chickens home. The gravy, on the other hand, may or may not be made that day. If I'm too busy, the giblets are gathered in a plastic bag and frozen for use within the next few weeks.

By the way, when you are preparing poultry for the freezer, do be sure to have a bowl handy in which to save the fat you remove from plump hens. Rendered slowly over low heat, strained and put in glass jars or plastic containers, chicken fat is a rich, savory shortening which can be used for great flavor dividends in baking bread, cakes and cookies, for frying, or as seasoning for soups and gravies. This is another example of the economies made possible by owning a home freezer, for the price of chicken fat in most areas is sky-high.

Freezing Turkeys

Like chickens, these may be purchased alive, dressed or eviscerated. They are classified according to age.

Young hens and toms are usually less than a year old. They are soft-meated and have flexible breast bones.

Old hens and toms are mature, more than a year old, with somewhat toughened flesh and hardened breast bones.

Cocks may be any age or weight, and are distinguishable by having darkened and toughened flesh.

Buying holiday dinner turkeys from a neighboring turkey farm during the seasons of the year when they do not command the high prices of Thanksgiving time is an economical practice, for they may be stored in your freezer up to twelve months.

Because these big birds take up so much valuable room in the freezer, however, you may not want to store more than a couple of them whole for your Thanksgiving and Christmas dinners. If their off-season price is so low that you cannot resist the temptation of buying several, a few of the batch may be cut into parts, as for fryer chickens, and frozen for casserole dishes, or into halves for company meals when only a few guests are present.

Others may be roasted and cut into sandwich slices, diced for turkey à la king or packaged in cooked serving pieces along with patties of dressing, with extra gravy in small glass or plastic containers. These frozen fast meals come in very handy on busy days or when company drops in.

Freezing Ducks and Geese

The young of both of these may be recognized by their soft flesh, somewhat pliable bills and easily dented windpipes. Feathers may have a rather downy appearance, too. Older birds have toughened skin and hardened bills and windpipes, and the feathers are fully developed.

POULTRY IN YOUR FREEZER

Be especially careful when drawing ducks and geese for freezing to remove the oil glands, which are useful to these birds in habitat but decidedly undesirable on your table. As with all poultry, chill them in the refrigerator before packaging them for the freezer in laminated freezer paper, cellophane protected by stockinette, or in cellophane and then in polyethylene bags.

RECOMMENDED MAXIMUM STORAGE PERIODS FOR POULTRY

Type	*Months at 0°F.*	*Months at 5°F.*
Chicken		
Whole	8-12	6-8
Parts	6-10	4-6
Sliced or diced	4-6	2-4
Livers	2-4	1-2
Stuffed roasting *	1-2	½-1
Turkey		
Whole	8-12	6-8
Parts	6-10	4-6
Sliced or cubed	4-6	2-4
Livers	2-4	1-2
Stuffed roasting *	1-2	½-1
Ducks, geese	6-10	3-4
Pre-cooked poultry	3	1

* Use moderate amount of spices in stuffing.

BULLETINS

The following pamphlets and bulletins are available for further study:

From the Superintendent of Documents, U.S. Government Printing Office, Washington, D. C.:
#G70–Home Freezing of Poultry (15¢)

Write also to the College of Agriculture in your state for additional pamphlets. Address your inquiry to the Department of Poultry Husbandry, (name of state) University.

13. Freezing Fish and Shellfish

If you live near an ocean, a lake or a stream and have a fisherman in the family, your home freezer can be the repository for quantities of delicious, nutritious, vitamin-rich protein food most months of the year—and it's all *free.*

This, at any rate, is an earnest argument I have heard given to wives who object to being left alone on week-ends during the fishing seasons while their husbands go traipsing off in imitation of so many Izaak Waltons. If she is wise, a wife rarely points out the fallacy of the word "free" while itemizing and calculating the cost of such appurtenances as fancy trout flies, rods and reels, minnow boxes and bottles of beer.

If it is brought straightaway to your kitchen from its watery home, the fish you freeze need not be tested for freshness. Just be sure the triumphant angler has kept it on ice; then chill it further in your refrigerator, and clean, package and freeze it as quickly as you can.

Buying fish in large quantities at economy prices in a store is another story entirely, however, and because this food is exceptionally perishable it is well to know how to determine its quality.

Three of your five senses will tell you at once whether a fish is fresh or stale:

Look first at the eyes, gills, belly walls, muscle tissue and vent of the fish. The eyes of *fresh fish* are bright and unwrinkled; the gills are bright red and covered with clear slime; the belly walls are intact; the muscle tissue is white; the vent is pink or tinged with pink, and does not protrude. *Stale fish* are likely to have dull, wrinkled and sunken eyes; the gills are dull brown or gray and the slime is cloudy; belly walls may be ruptured, exposing the viscera; muscle tissue is pinkish, especially around the backbone; the vent is brown and protruding.

Next, smell the fish. Be brave. If it is fresh, it will smell very fishy—almost like a whiff of the sea (if salt water fish) or of weeds (if fresh water fish). If it is stale, there will be an unpleasantly sour or putrid odor.

Last, touch the fish with the ends of your fingers. A fresh fish will have firm flesh. In very fresh fish, such as those newly caught, the bodies will be quite stiff. Your fingers should make no dents. Soft, flabby flesh characterizes stale fish, and impressions made by your fingers will remain.

If your household is suddenly showered with a rain of fish, prepare only a few at a time and keep the rest on ice or in the refrigerator. They should be cleaned, scaled, washed and prepared as if for cooking before they are packaged and frozen.

If your husband is off on a fishing trip lasting several days, it is to be hoped that he will clean them and store them in a cold place at the lodge or camp, or in the boat's ice chest if he is deep-sea fishing. When he brings them home, he will probably have them packed in ice, keeping them off the bottom of the container by means of a rack so that they won't become bloated with water.

How to Clean Fish

Wash thoroughly in cold salted water—1 tablespoonful of salt for each quart of water used.

Lay the fish flat on a board, holding firmly to prevent its slipping.

Using a special fish scaler or a sharp knife, scrape scales from tail toward the head.

On the belly side of the fish, make an incision which runs the entire length of the body, from vent to head. Remove entrails carefully.

Remove fins. Cut head off above the collar bone and through the backbone. Remove tail.

Wash again quickly in salt water as before, removing any loose membranes and blood which may remain.

After it is cleaned in this manner, fish may be frozen whole or cut into steaks and fillets. Very large fish, of course, are preferably cut into steaks—except the one your husband is so proud of he won't hear of its being frozen or served any way but whole, as proof that it didn't get away.

Freezing a Whole Large Fish

First, put the cleaned and washed fish in the freezer without any wrappings, placing it against the side wall of a chest freezer, in a fast-freezing compartment if you have one, or directly in contact with a freezer plate shelf of an upright. Have a large pan of extremely cold water handy, and when the fish is frozen dip it in the water to form a thin glazing of ice.

Return the fish to the freezer for an hour, then repeat the process of dipping and freezing until the ice glaze has built up to a thickness of ⅙ inch to ¼ inch.

The fish may then be stored without wrappings if it is going to be used within a period of two or three weeks.

If you are uncertain about when you will serve it but want it to remain unwrapped for the sheer pleasure of gloating over it from time to time, take it out of the freezer after three weeks and renew the glaze, some of which will evaporate in the cabinet. Or, to be on the safe side altogether, wrap the ice-coated fish in laminated freezer paper.

Freezing Small Fish, Steaks and Pieces, and Fillets

Divide fish, pieces or fillets into meal-sized portions, separating them with double thicknesses of cellophane. Wrap a single meal's worth very carefully in a strong, moisture-vapor-proof material such as heavy aluminum foil or laminated freezer paper. Be sure to exclude all air, seal tightly, label and freeze.

IN A BRINE DIP

This method is recommended for the lean varieties of fish only, as salt has a tendency to hasten oxidation, thus leading to rancidity. Do not, therefore, brine-dip any of the high-fat content fish such as albacore, bonito, butterfish, eels, grayfish, halibut, herring, mackerel, lake trout, pilchard, pompano or salmon.

To make a brine dip, mix a solution of ½ cup salt to each quart of water used. Dip the fish into this solution for 30 seconds. Drain, then wrap and seal as above. Fish frozen after brine-dipping should be used within three months.

IN AN ACID DIP

There are a number of good ascorbic-citric acid preparations on the market, sold under various trade names. One of these is called "A.C.M." and is put out by Chas. Pfizer & Co., Inc., the chemists who also manufacture this country's supply of the wonder drug, terramycin. Ascorbic acid is an effective anti-oxidant, combating the development of rancidity in fish, which contain varying percentages of unsaturated fatty acids which are especially vulnerable to off-flavors and odors caused by reaction to atmospheric oxygen. Citric acid has been used for many years by packers in the fish industry to preserve the color and flavor of their products.

Make a 1 per cent solution of ascorbic-citric acid (one part powder to 100 parts of water) or follow the manufacturer's directions for the acid preparation you are in the habit of using. Dip fish into this solution for 30 seconds, drain, wrap and seal.

IN ICE BLOCK

Several small fish, steaks or fillets may be placed in refrigerator trays or in loaf pans, covered with water and frozen. Separate fish or pieces with double strips of cellophane for easier handling when they are removed from the freezer. When the blocks are solidly frozen they may be removed from the pan, wrapped in cellophane or freezer paper and stored.

You may, if you prefer, put fish in quart- or half-gallon-size waxed tub-type containers and cover with cold water to within a half inch of the containers' tops. Be sure all of the fish are completely covered. If necessary, crumple a piece of cellophane over the top of the fish before covering the container.

RECOMMENDED MAXIMUM STORAGE PERIODS FOR FISH

Varieties	*Months at 0°F.*	*Months at 5°F.*
Lean		
Abalone, black or white bass, cod, red, flounder, grouper, haddock, hake, king, yellow perch, pike, pollock, porgy, red snapper, shellfish	8-12	4-6
Fatty		
Albacore, alewife, barracuda, blue, bonito, butter, carp, croaker, eels, gray, halibut, herring, mackerel, whiting, lake trout, mullet, white perch, pilchard, pompano, salmon, shad, striped bass	6-8	2-4

Freezing Shellfish

These, the fruits of the sea, are happily at home in a freezer and provide many delectable, luxurious meals throughout the year, even at those times when they are fabulously expensive in the markets or not available at all.

Oysters are always in season when you have a home freezer. The traditional R need not concern you at all, for you can freeze them in December and enjoy them in May.

Maine lobsters are now being packaged along with their native seaweed and shipped alive to any part of the country.

The frozen food industry has been influential in increasing the shrimping activities in the Gulf States, and more and more fresh shrimp are finding their way into local markets. Summer clams, too, are being shipped by the bushel in refrigerated cars to many inland communities never before able to purchase them for home consumption.

These delicious and wonderfully nutritious foods (they are chock full of essential vitamins, minerals and proteins) are, in my opinion, well worth the extra time and trouble it takes to prepare them properly and freeze them in quantity.

CRABS, HARD SHELL

Boil for 15 or 20 minutes in a quantity of salted water. Cool slightly under running cold water before picking out the meat. Flake or lump the body meat, removing all membranes, and keep claw meat whole and separate. Pack crabmeat tightly in plastic or waxed tub containers, or in glass jars. Loosely packed crabmeat freezes more slowly and may cause toughening. Seal packages securely before freezing.

CRABS, SOFT SHELL

Clean soft shell crabs by removing the spongy substance found underneath the tapering points on both sides of the back; then, from the front part of the shell remove the apron, a small piece at the lower part which ends in a point. They may be boiled for 20 minutes, cooled, wrapped individually and frozen, or they may be frozen after cooking by your favorite recipe.

CRAYFISH OR LOBSTERS

Boil for 20 minutes in salted water, or split and broil under a medium flame. Cool. If you have plenty of room in your freezer, lobsters or tails may be frozen in their colorful shells for a short period, each one wrapped separately in aluminum foil or laminated freezer paper. If you are crowded for space, pick out the edible meat and follow the instructions given for packaging crabmeat.

CLAMS

Clean thoroughly, either by brushing shells briskly or by scrubbing them under running water. Open with a knife, or allow them to open themselves by putting them on the bottom of a large pan over low heat. Remove all dark material from the clam meat. Strain the clam juice until it is free of sand.

Put clams in glass, plastic or wax-impregnated containers and cover them with their own juice. If there is not sufficient juice to cover them completely, add enough water to do so. Care must be exercised to cover the clams entirely with water, because any which are exposed to air space in the top of the container may become discolored. Allow at least ½-inch expansion space, close and seal the containers securely.

OYSTERS

Wash in salt water, and open carefully with an oyster knife, holding the oysters over a bowl to preserve the juice. From this point, prepare for freezing according to the directions for clams.

SHRIMP

These may be frozen either raw or cooked, but cooked shrimp have a tendency to toughen in the freezer. Beheaded, they may be frozen either in or out of their shells. I prefer to freeze them shucked and deveined, packaged tightly in rigid moisture-proof containers. If you prefer to cook them first, you can prevent some of the toughening by simmering them only long enough to turn them slightly rosy—about five minutes—finishing the cooking process when you remove them from the freezer for serving.

SCALLOPS

Shucked scallops are frozen in the same manner as clams or oysters.

Recommended Storage Periods for Shellfish

Because of their extremely low fat content, all types of shellfish may be stored at 0°F. *when properly prepared and packaged* up to 10 months; at 5°F. up to six months. Precooked shrimps will have better texture and flavor if not stored for longer than three months.

14. Freezing Wild Game

As civilization progresses, our crazy, wonderful world seems to take an occasional turn around an imaginary second axis, coming face to face again with old, long out-moded customs in the new guise of modernity. We keep going back to the old ways, even if we do take a jet plane to get there. We preen ourselves on our great inventions of telephonic and wireless communication, for they permit us to talk to each other, without yelling, at distances greater than across the dinner table. Yet how different in essence, if not in paraphernalia, are these modern devices from the smoke signals and tom-toms of ancient or less civilized peoples?

Once, before food came in packages, the hunter was an honored and important personage in the community, for the tribe's food supply depended on his skill and cunning in the forest. As the world grew older and wearier, hunting became less essential work than vigorous play; the hunter may by other standards have been an important personage in his community, but he tracked and killed animals for the sport of it and not because his family and neighbors needed to eat.

Our pioneering forefathers held game animals in high esteem, for they had to depend on wild animals tracked down in field and forest for their meat supply, and were usually careful not to kill more than were needed to feed their families and communities.

After frontiers were settled and food depots were centralized, men continued to hunt for pleasure without regard to conservation of the country's regional or migratory wildlife. Game laws were lax or nonexistent, and our national wildlife was in peril of becoming extinct. Federal and state laws were enacted, some of them so stringent that in a few states today certain animals and birds may not be legally taken at all.

Within the limitations imposed by their states, today's huntsmen eat what they want of their catch or kill, frequently still in their boots and sitting happily around a campfire. They share a certain portion of their bag with less outdoorsy friends and neighbors. And that newest of contributions to strictly modern living, the home freezer, is returning to the hunt its original and moral purpose—food supply.

Brought home in triumph, large and small game animals and birds are processed, prepared and wrapped for freezing. Long after the hunting season has closed, a succulent venison steak or a tenderly basted quail is enjoyed not only for its gourmet qualities but also for the fine memories it evokes and the splendid opportunities presented for a little masculine bragging.

FREEZING WILD GAME

Home freezers as well as community frozen food locker plants are currently being put to use for more than just deer and duck by hunters and their families who know the special savor of many species of wildlife. A glance through sporting magazines will reveal that in many parts of the country, freezers are dedicated to the storage of such exotic edibles as prairie dog, woodchuck, wild raccoons, beaver, muskrat, skunk, armadillo, antelope, moose, reindeer, bear, boar, wild turkey, quail, pheasant, grouse, partridge, prairie chicken, sparrow, crow and woodcock.

Rattlesnake meat, too, is considered by many gastronomes to be a rare delicacy, and at least one company puts this up in colorfully labeled tins which are sold in some of the country's most exclusive markets. It can be frozen, but I must be excused for not having a recipe handy.

Handling Game Birds for Freezing

Although some authorities claim that game birds may be prepared for freezing exactly as though they were domestic poultry, I think a few extra precautions are wise if you expect an optimum result.

The excessively strong, gamy flavor of pheasant, duck and other birds usually comes from the food the birds were eating just before they were shot. Domestic fowl are put on an enforced diet prior to their demise, food being withheld for at least from eighteen to twenty-four hours before they are killed. This empties the crops. Such a measure is obviously impossible with elusive game birds, however, and so the next-best thing is to eviscerate them immediately after they have been brought down, preventing the strong-tasting food in their crops from permeating the flesh.

Newly shot birds should not be stuffed into waterproof pockets of hunting jackets because this does not give them a chance to cool.

If thorough drawing is inconvenient in the field, the birds should at least have their craws and intestines removed until brought home, when a more complete cleaning job can be done. If you are uncertain about game birds' field care, a recommended flavor-saver is to soak them overnight, after they have been drawn and plucked, in a mild acetic acid solution or in a two-gallon vessel of water to which has been added ½ cup of white vinegar.

While skinning is the quickest way to remove feathers, it is better to pluck the birds unless you are prepared to eat dry meat when cooked.

A good method to draw the blood from game birds is to soak them in a brine solution for several hours (2 tablespoonfuls of salt to each quart of water used). Rinse well and wipe dry before proceeding with freezing.

Two opposing epicurean schools of thought have waged a long and stormy controversy over the years—to hang or not to hang. Some people believe that mild flavor is preferred, and get their birds into the freezer as quickly as possible. The opposite school scorns this as sissy stuff, declaring that birds should be hung in a cool place until they are almost strong enough to flap their wings and fly off. On this subject I must remain neutral, for game bird gourmets are partisan and incontrovertible. I merely reserve my right to accept dinner invitations only from the sissy group.

After birds have been drawn, plucked, hung (or not) and washed very carefully inside and out, put them in the refrigerator to chill thoroughly. When they are cold, treat them as you do chickens for the freezer. They may be frozen whole for roasting or cut into parts and packaged in meal-size portions for frying or casseroles. Package in plastic bags, being sure to exclude all air, or in laminated freezer paper or aluminum foil protected with stockinette. Freeze promptly.

Handling Rabbits, Squirrels, Woodchucks and Other Small Game for Freezing

Small game animals should be beheaded, eviscerated and skinned as soon as possible after they are bagged. Particular care should be taken in cleaning out the shot wound.

With rabbits and all other four-legged game, the hunter should carefully examine the liver, lungs and intestines to be sure there are no white spots or unusual growths of any kind. When these appear, the animal may not be quite safe to eat. Particular care should be taken with wild rabbits, which may transmit the disease *tularemia.* Rabbits should be handled with rubber gloves, for the disease is transmitted through lesions, entering the handler's bloodstream with serious, sometimes fatal, results. Cooking rabbit meat in boiling water for 30 minutes, or until thoroughly well done in a hot oven, will destroy the microorganisms.

Small game carcasses can be bled in the same way as game birds, by soaking for three hours in a brine solution of 8 tablespoonfuls of salt to each gallon of water used. Rinse thoroughly and dry. Disjoint, cut into parts, and package according to the directions for cut-up chickens.

Handling Big Game for Freezing

Many frozen food locker plants located in hunting-ground states offer a valuable service to sportsmen and their families. They will skin, butcher, age, cut, wrap, mark and freeze the meat for a nominal charge.

If you avail yourself of the services of such a plant, you might discuss the routine with the manager, keeping an important factor in mind. Plant operators have found that they can work more quickly and conveniently with venison by quick-freezing the loins and legs before cutting them into steaks and roasts, as this meat is flabby and soft and free from fat, therefore difficult to cut into even slices. The best venison, however, comes from deer, moose or elk which have been hung in a cooler for a period of time up to a month. Very fresh, unhung deer meat is likely to be even tougher than un-aged beef, which has some fat surfaces and marbling to lend tenderness.

During the cooling period, enzymic action leavens and partially pre-digests the meat. Of course, if a deer is allowed to hang *too* long, our old friends the enzymes will continue their action until the meat becomes spoiled.

If you want your venison to be on the tender side, arrange with your locker plant manager to hang the deer for firming rather than to pre-freeze it for easy cutting.

The most succulent and pleasant-flavored venison comes from deer which are given prompt and careful attention as soon as they are killed—in other words, while the hunter is still in the woods. Objectionally strong flavors are due almost entirely to delay or inadequate dressing, insufficient bleeding, or to failure in cooling properly.

Field care includes eviscerating the deer immediately, using a sharp knife and a practiced or guided hand. Wrap the liver separately and return it to the cavity.

The rectal organs should be cut out and thrown away at once. Wash out the cavity and take the time to dry it thoroughly. When driving home, keep the cavity exposed to air. A clean branch may be used to keep the cavity open.

Whether you do your own cutting or hand the deer over to a butcher or local locker plant operator for this purpose, retain only the more desirable cuts as steaks and roasts, to be frozen for venison feasts at some later date. The beef chart on page 137 can be followed as a guide for cutting deer.

Some locker plants will grind neck, shoulders, shanks and other parts for you, but most of them are too busy during the deer season to perform this service. Take these parts home with you and put them through your own food chopper for delicious deerburger. Before freezing it in meal-proportioned packages, chill the ground meat overnight in the refrigerator. Do not add salt at this time, although other seasonings such as pepper and grated onion may be worked in, if desired.

The meat of herbivorous animals, such as deer, is likely to dry out in storage because of its limited fat content. It is wise, therefore, to have the steaks and roasts cut into the sizes you will later want for your recipes so that they can be easily marinated while defrosting to safeguard further against possible toughness.

Game Laws Relating to Freezing

The length of time you may keep game birds and other wildlife in your home freezer is controlled by national and state laws, and many states require licensed hunters to obtain permits for sub-zero storage, whether in community locker plants or in domestic cabinets.

Most of the state laws were passed before home freezing became popular and widespread, and many of the sovereignties have since revised or amended their legislation to take cold-storage possibilities into consideration. Those states which have not yet got around to appropriate new legislation will, on application, extend the limits indicated in the Table by issuing special permits.

Check carefully the local laws governing storage as well as bag and possession limits, and be sure to label all game packages with accurate information as to the dates the animals were taken and frozen. It is true that the storage laws are difficult to enforce. They do exist, however, principally for your own protection, although the lean quality of most game birds and animals makes them particularly adaptable to long-term freezer storage.

You will note from the chart that even with permit exten-

STATE LAWS RELATING TO FREEZING

(Except Migratory Waterfowl and Game Birds)

State	*Permit Regulations*	*Amount of Game Which May Be Frozen*	*Length of Storage Time Permitted*	*Officials from Whom Copies of Game Laws May Be Obtained*
Alabama	No permit required.	All legally taken.	90 days after the close of the season.	Director, Dept. of Conservation, Montgomery 4.
Arizona	No permit required for game kept less than 60 days following end of open season for species. Permit issued on request for longer storage, 50¢ for each legal bag limit.	One legal bag limit of each species.	Under special permit, no maximum time limit.	Director, Game and Fish Commission, Arizona State Bldg., Phoenix.
Arkansas	No permit required, but stored game must bear tags showing kind, number, date, license number and name of owner.	All legally taken.	Indefinite period.	Executive Secretary, Game and Fish Commission, Little Rock.
California	No permit required.	Possession limit.	Indefinite period.	Executive Officer, Fish and Game Commission, Ferry Bldg., San Francisco 11, or from Dept. of Fish and Game, Sacramento 14.

Colorado	Permit attached to license.	Possession limit.	Not to exceed 30 days prior to the opening of next year's season for species.	Director, Game and Fish Commission, Denver 5.
Connecticut	No permit required.	Bag limit for each species.	Indefinite period.	Superintendent, State Board of Fisheries and Game, Hartford 1.
Delaware	No permit required.	Bag limit.	5 days after close of season. No provision made for freezer owners.	Board of Game and Fish Commission, Dover.
Florida	No permit required.	All legally taken.	Indefinite period.	Director, Game and Fresh Water Fish Commission, Tallahassee.
Georgia	For storage longer than 5 days, free permit issued.	All legally taken.	With permit, maximum of 90 days.	Director, Game and Fish Commission, 412 State Capitol, Atlanta.
Idaho	No permit required.	Possession limit.	Indefinite period.	Director, Dept. of Fish and Game, Boise.
Illinois	No permit required.	All legally taken.	Indefinite period.	Director, Dept. of Conservation, Springfield.
Indiana	Free permit required, issued on application.	Legal bag or possession limit.	Six months from close of open season of each species.	Director, Division of Fish and Game, Conservation Dept., Indianapolis 5.
Iowa	Permit required for game held longer than 10 days beyond season. No charge.	Lawful possession limit.	Until June 30 following close of season for each species.	Director, State Conservation Commission, Des Moines 9.

STATE LAWS RELATING TO FREEZING (*Continued*)

State	*Permit Regulations*	*Amount of Game Which May Be Frozen*	*Length of Storage Time Permitted*	*Officials from Whom Copies of Game Laws May Be Obtained*
Kansas	No permit required.	Two days' bag limit of in-State game animals and birds.	30 days following close of season, except for out-of-State big game legally killed, which may be held in storage in accordance with individual state law.	Director, Forestry, Game and Fish Commission, Pratt.
Kentucky	No permit required.	Two days' possession limit. Doves, one day's limit.	90 days following close of season.	Dept. of Fish and Wildlife Re sources, Frankfort.
Louisiana	Permit required, issued free on application.	All legally taken.	30 days before next open season on resident game.	Director, Dept. of Wild Life and Fisheries, New Orleans 16.
Maine	No permit required.	All legally taken.	No restrictions.	Chief Warden, Inland Fisheries and Game, Augusta.
Maryland	Permit required for game kept longer than 5 days after close of season, except deer, which may be kept 30 days.	Possession limit.	Not to exceed 180 days from date of kill.	Director, Game and Inland Fish Commission, Baltimore 2.
Massachusetts	No permit required.	Possession limit, all legally taken.	Indefinite period.	Dept. of Conservation, Boston 8.

Michigan	Permit required, issued free on proof of lawful possession, for game held longer than 60 days after close of season.	All lawfully taken.	With permit, for an indefinite period.	Director, Dept. of Conservation, Lansing 13.
Minnesota	No permit required.	Possession limit.	No restrictions.	Director, Division of Fish and Game, St. Paul 1.
Mississippi	No permit required.	All lawfully taken.	Indefinite period.	Director, Game and Fish Commission, Jackson.
Missouri	No permit required.	All lawfully taken.	Big game (except deer) and upland birds until July 1 of each year. Deer, rabbits, frogs and furbearing animals not to exceed 90 days following close of season for species.	Director, Conservation Commission, Jefferson City.
Montana	No permit required.	One legal limit.	Indefinite period, but no additional animals may be taken while in possession of any part of same animal.	Secretary, Fish and Game Commission, Helena.
Nebraska	No permit required.	All lawfully taken. Must be reported to Game Commission.	90 days following close of season for each species.	Game, Forestation and Parks Commission, Lincoln 9.

STATE LAWS RELATING TO FREEZING (Continued)

State	*Permit Regulations*	*Amount of Game Which May Be Frozen*	*Length of Storage Time Permitted*	*Officials from Whom Copies of Game Laws May Be Obtained*
Nevada	No permit required.	One possession limit.	Indefinite period.	Director, State Fish and Game Commission, Reno.
New Hampshire	No permit required.	Two days' lawful limit, all game except deer, limited to one carcass.	Indefinite period.	Director, Fish and Game Dept., Concord.
New Jersey	Must be tagged and dated if held in public freezer. No permit required for home freezer.	All legally taken.	No limit.	Commissioner, Dept. of Conservation and Economic Development, Trenton 7.
New Mexico	No permit required.	All legally taken.	To March 31 next after close of season.	Dept. of Fish and Game, Santa Fe.
New York	Free permit required for possession longer than 60 days, issued on application. Deer and bear may be held until March 1 following close of season.	All legally taken.	Permit may be renewed on application.	Commissioner, Conservation Dept., Albany 7.
North Carolina	No permit required.	All legally taken.	Indefinite period, if declared to wildlife protector.	Executive Director, Wildlife Resources Commission, Raleigh.
North Dakota	No permit required.	Possession limit.	Until Sept. 1 following close of season.	Commissioner, Game and Fish Dept., Bismarck.

Ohio	No permit required.	Possession limit.	Indefinite period.	Director, Dept. of Natural Resources, Columbus 12.
Oklahoma	No permit required.	All legally taken.	Indefinite period.	Director, Game and Fish Dept., Oklahoma City 5.
Oregon	No permit required. Elk and deer must be tagged.	All legally taken.	Indefinite period.	State Game Commission, Portland 8.
Pennsylvania	No permit required.	Season limit.	To July 2 following close of season.	Director, Game Commission, Harrisburg.
Rhode Island	No permit required.	All legally taken.	No limit.	Administrator, Division of Fish and Game, Providence 2.
South Carolina	No permit required; packages must be marked with number of license.	All legally taken.	Unlimited time.	Director, Wildlife Resources Dept., Columbia.
South Dakota	No permit required.	All legally taken.	No limit.	Director, Dept. of Game, Fish and Parks, Pierre.
Tennessee	No permit required.	Two days' limit.	Indefinite period.	Director, Game and Fish Commission, Nashville 10.
Texas	No permit required.	All legally taken.	No time limit.	Director, Game and Fish Commission, Austin.
Utah	No permit required.	All legally taken.	No time limit.	Director, Fish and Game Commission, Salt Lake City.
Vermont	No permit required.	Legal limit.	Indefinite period.	Director, Fish and Game Service, Montpelier.
Virginia	No permit required.	All legally taken.	Indefinite period.	Executive Director, Commission of Game and Inland Fisheries, Richmond 13.

STATE LAWS RELATING TO FREEZING (*Continued*)

State	*Permit Regulations*	*Amount of Game Which May Be Frozen*	*Length of Storage Time Permitted*	*Officials from Whom Copies of Game Laws May Be Obtained*
Washington	No permit required.	All legally taken.	To August 1 following close of season.	Director, Dept. of Game, Seattle 9.
West Virginia	No permit required, except for deer or bear meat.	Not to exceed daily or season bag limit, whichever is the smaller number.	60 days after open season. Deer and bear meat for additional 4 months, with permit.	Director, Conservation Commission, Charleston.
Wisconsin	No permit required.	All legally taken.	To June 30 following close of season, except deer, which may be stored (properly tagged) until consumed.	Director, Conservation Dept., Madison 2.
Wyoming	No permit required.	Possession limit.	Until consumed.	Game Warden, Game and Fish Commission, Cheyenne.

sions some states still limit freezer storage of legally taken game to as little as two or three months after the season has closed, while others permit you to keep the game meat indefinitely, granting you the privilege of deciding whether or not it is still edible. The laws of the more restrictive states will undoubtedly be changed as constituents convince their legislators of the freezer's logical and valuable contribution to real conservation, both of wildlife and of food.

NATIONAL GAME LAWS

The federal regulations take precedence over any and all state laws in controlling the possession and storage limits of *migratory waterfowl,* including wild ducks, geese, pigeons, doves, etc. According to the United States Department of the Interior, Fish and Wildlife Service, such migratory waterfowl may be legally possessed for a period *not to exceed 90 days following the close of the regular season.* Federal regulations further provide that the feet, head and head plumage shall remain on the birds while being possessed in storage.

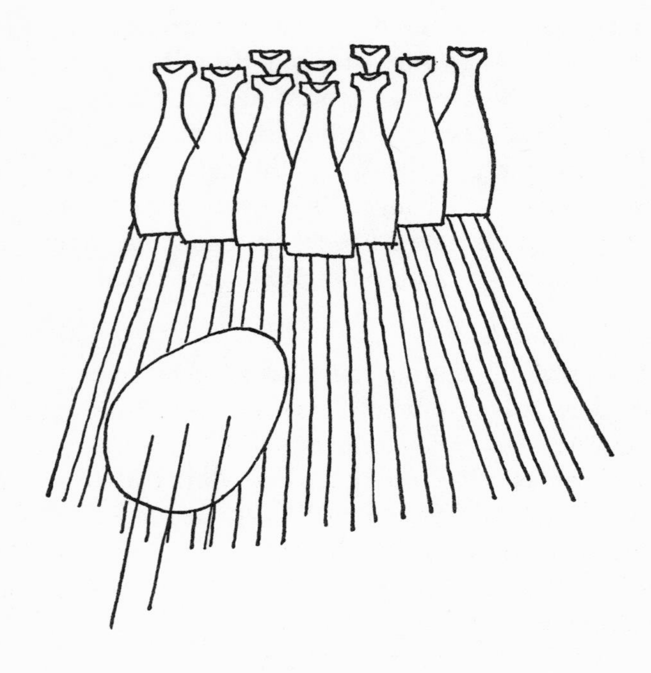

15. Freezing Dairy Foods

"Country convenience for city folks." That's what your home freezer can give you if you are city folks without benefit of cows, laying hens, butter churns and cream separators.

One of the great economy features of freezer ownership that you will value and appreciate is your freezer's ability to preserve successfully such costly perishables as eggs, cheese, butter and heavy cream. And, of course, the children usually cannot be convinced that freezers were invented for any reason other than ice cream. Home-made or store-bought, it makes no difference, just so long as there's plenty of it in as many of their favorite flavors as Mother can be induced to keep within easy reach.

Eggs, especially, fare well in the freezer both from the standpoint of economy and convenience. This superb protein food is so variable in price throughout the year that it is good practice to watch your local markets for the times of the year when egg prices drop, and buy up large quantities for freezing.

In the section of the country where I live, egg prices take a swooping downward plunge in the spring of the year and soar to almost double their springtime cost in the winter. By glancing at daily livestock market quotations and noticing the trend, and by watching the changing signs in my neighborhood supermarkets where I shop for unfreezables, I get a pretty good idea of when to pounce on a large egg haul against the days of inflation.

One year, for example, I froze a gross of grade A white and brown eggs for which I paid 45 cents a dozen and used more dozens of fresh eggs while the price remained low. When winter rolled around and the local retail price rose to 90 cents a dozen, I used my 45-cent fresh-frozen eggs with a profligate hand and an unbecoming sense of smug self-approval.

Whenever it is possible, of course, I drive out into the country and watch for roadside signs that announce "strictly fresh farm eggs," turning into a likely driveway to come to terms with the farmer.

If you follow this pleasant adventure of egg-hunting, keep in mind that heat is no friend to an egg after it has left its mother's warmth.

The farmer who gets my business is not the one who arranges nice, clean eggs in geometric patterns on a sunny roadstand. My egg farmer is one who is satisfied with a hand-lettered sign announcing his produce, with an arrow pointing to where the eggs are. His eggs are kept unwashed and refrigerated until I come along.

Nature has provided a perfect package for eggs, but the shell has thousand of tiny little pores which, in a freshly laid eggs, are protected by a film-like covering. This covering disappears after a while even if left alone; washing it removes it too soon and a form of dehydration begins to take place within the egg. Some egg authorities have recently reversed this thinking. They say, now, that it is all right to wash eggs. The earlier theory makes more sense to me, however.

If left in the sun or in a warm room, eggs are subject to further deterioration. The albumen (white part) becomes thinner and weaker, and soon it is just too tired to be able to hold the yolk nice and round in its central position.

I buy white or brown eggs without discrimination, for the color of the shell is influenced only by the breed of the hen which produced the egg. There is absolutely no difference in flavor, grade or nutritional value. I have had dark-brown eggs whose yolks were as pale as straw, and pure white ones whose yolks reflected the coppery tones of a western sunset. The color of the yolk is influenced only by the type of diet which was fed to the hens while they were producing.

Freezing Eggs

I freeze eggs in a number of ways and find all of them useful. Once, in complete disregard for all the rule books which say, "Never freeze whole eggs in their shells because expansion will cause the shells to crack," I froze whole eggs in their shells. My logic was that no wrapping material I could buy would be superior to nature's own tamper-proof package.

When, in due course, the shells burst open on my freezer plate shelves, I sealed up the cracks with scotch tape and said, "There!" They were not a total loss, but the results were so unsatisfactory that I was glad my trail-blazing zeal had been limited to experimenting with only a half-dozen eggs.

I tried defrosting one at room temperature, meaning to poach it. It was edible, but the white was tough and rubbery, and the yolk gummy. I tried hard-cooking another without first defrosting it, and this was a little better, but not much.

I finally defrosted the remaining four and hard-cooked them for use as a chopped garnish for spinach. Nothing was wasted, thank goodness, but the experience marked my first and only real freezer failure.

Select for freezing only very fresh eggs of good grade, and wash them. Break each one into a saucer to be sure of its quality. It should be odorless and the yolk, well centered in the albumen, should not be streaked with blood.

WHOLE EGGS

For single eggs, I stir together a dozen or so, lightly, with a fork—firmly enough to mix the yolk and white but not enough to beat any air in. I pour the stirred eggs into my refrigerator ice-cube trays and freeze them immediately. Sometimes, if my trays are busy elsewhere, I use waxed paper muffin cups nested in a muffin tray.

When the eggs are solidly frozen as cubes, I pop them out of the trays and store them in laminated bags. If I have used the muffin cups, I store them in their little paper dresses, also in laminated bags. One cube or muffin is the equivalent of one whole egg. This same system can be used for separate egg yolks or whites, too.

Inasmuch as many of my recipes call for varying numbers of whole eggs, I also freeze different quantities in tub-type containers, small laminated bags and cellophane-lined pint boxes.

EGG WHITES

These need no special preparation, and they are every bit as good as fresh egg whites for baking or meringues. Merely separate the eggs (I use a 10-cent plastic egg separator for this) and package the whites in the amounts called for in recipes you habitually use.

EGG YOLKS

Because egg yolks coagulate for some reason while in freezer storage, they must be mixed with something else to prevent gumminess or an undesirable rubbery texture in the thawed product. What you use as this preventive factor will depend on your future recipes. You may use salt, sugar, honey or corn syrup, in the following proportions:

Salt—½ teaspoonful to 6 egg yolks (or 8 whole eggs)
Sugar—2 teaspoonsfuls to 6 egg yolks (or 8 whole eggs)
Honey or syrup—1 tablespoonful to 6 egg yolks (or 8 whole eggs)

Stir the mixtures very lightly with a fork or wire whisk, trying not to mix much air in. The egg yolks must *not* be fluffy or lemon-colored at this time. Package in quantities desired in moisture-vapor-proof containers. Be sure to label all your frozen egg containers fully and accurately. For example:

Cubes—whole eggs; egg yolks; egg whites
Containers—6 whole eggs; 4 egg whites; 2 egg yolks, salt; 4 egg yolks, sugar, etc.

Eggs are liquid, so when you package them be sure to allow from ¼ to ½ inch at top of the containers for expansion. Date all packages.

Eggs may be stored in your home freezer for as long as a year at 0°F., or up to six months at 5°F. Recently I "found" a package of whole eggs in the back part of my freezer's bottom shelf which was dated two years earlier. They were fine. You will ordinarily find it best, however, to use up your frozen eggs by the time the next low-price season comes along.

When you are ready to use some frozen eggs, take them from the freezer and defrost them in their sealed containers in the refrigerator for best results. If you are in a hurry and forgot to take them out of the freezer in time, they may be defrosted at room temperature or in a bowl of cold water (unopened, of course). Use defrosted eggs immediately, as their quality deteriorates if allowed to stand.

Equivalents

1 tablespoonful defrosted egg yolk equals 1 egg yolk.
1½ tablespoonfuls defrosted egg white equals 1 egg white.
1 pint of defrosted whole egg equals 8 whole eggs.

Freezing Butter

The price variability of eggs applies also to butter, and here again your home freezer comes into welcome use as a means of real economy. If you buy large quantities of commercially packaged butter in rolls or in bars, you may leave it in its original wrapping but cover this with an outer-wrap of aluminum foil or heavy laminated freezer paper.

In some sections of the country, many of the dairies sell ten- and twenty-pound tubs of sweet or lightly salted butter to freezer owners at substantial savings. If you buy butter this way, it is best not to leave it in its original tub. Scoop it out and repackage it in smaller rounds or oblongs. While the butter is still fairly soft, pat it into round tubs (the pint size) or into small wax-impregnated boxes.

For double protection, either wrap the containers in aluminum foil or laminated paper or put them into large polyethylene bags. Take out of the freezer only as much butter as your family will consume within a week.

Salted butter will store safely in the freezer for from one to three months, unsalted from three to six months.

Freezing Cheese

Most types of cheese are excellent subjects for the freezer, although at times I have not been too happy with the fine-textured cream cheese, which seems to fall apart a little when it is thawed. Real economies can be effected by buying gallon-sized tins or containers of pasteurized cottage cheese or those great big wheels of Cheddar, repackaging them in sizes sufficient for a day's supply (if cottage cheese) or a week's (if Cheddar). A week's supply in my house may differ considerably from a week's supply in yours, however. Cheese disappears at a terrific rate of speed around here.

Don't attempt to freeze a large brick or wheel of cheese without slicing or cutting it into portions, for cheese has a tendency to crumble if it is cut while in the frozen state.

Cottage cheese goes nicely into those 16-ounce waxed tub containers or in plastic boxes of the same size. Cheddar and other hard types can be cut into half-pound wedges or sliced for sandwich spreads. Wrap wedges in plastic film and then in heavy laminated freezer paper. Be sure to slip double or folded thicknesses of cellophane between slices before wrapping and freezing them.

Cottage cheese keeps up to four months at 0°F. Defrost the containers unopened in the refrigerator before using.

Cheddar and other hard cheeses will keep up to six months. Defrost this also in its wrapper, preferably in the refrigerator.

Freezing Milk and Cream

In an emergency or in the event of epidemics, sweet pasteurized homogenized milk and buttermilk may be frozen satisfactorily and stored up to two or three months in your home freezer. Even without an emergency, milk for cooking purposes may be bought by the gallon or in even larger containers if there is a dairy in your neighborhood, and frozen in quarts, thereby effecting still another economy.

It has always seemed to me that preparing large batches of baby formula in advance and freezing it in sterile bottles would be a fine time-saver for busy mothers, but none of the experimental stations or milk companies I have contacted on this problem has committed itself as to whether this is or is not possible. However, the production manager of one of the country's biggest dairy companies confided to me that milk frozen at 0°F. or below can be held at 0°F. for as long as six months. This company has been shipping frozen whole milk to the armed forces for years.

When freezing milk, be sure to leave one-inch expansion space at the top of the freezer carton or jar, and package in containers sufficient for one-time use.

I have recently learned to use the packaged dry skim milk solids to great advantage in cooking and baking, and even for drinking. It is an economy and a convenience, and the absence of milk fat does my diet no harm. It is frequently a nuisance, however, to mix up a quart or so in the middle of an impulsive bread-baking spree, and so I've gotten into the habit of keeping a few quarts toward the front of my freezer, replacing the one I use with a new batch while I'm waiting for the bread to rise.

Heavy cream, too, keeps extremely well in the freezer. If it is pasteurized, it may be stored up to four months. Unpasteurized, it should be kept no longer than three or four weeks. The best freezing results are obtained from cream which is sweet and rich, containing at least 40 per cent butter fat by volume. Freeze enough in several containers for one-time use, as with milk. I put mine in tiny 8-ounce plastic containers, leaving ½-inch expansion room. Allow frozen cream to defrost unopened in the refrigerator, and give it a few brisk stirs before using or serving it.

My upright freezer comes into useful service for keeping whipped-cream toppings on hand, for I drop individual servings on cellophane and put it on one of the fast-freezing plate shelves. When the little mounds are frozen, I simply drop them into polyethylene bags and use them up within three or four weeks.

Ice Cream in the Freezer

Drug-store or dairy ice cream can be kept in economically purchased 2½-gallon, gallon and half-gallon containers in your home freezer, and will probably keep for as long as your family allows it to. About one month is the limit—by the freezer's standard, not the family's!

If you make your own, using your favorite recipe or one of those starting on page 367, pour it into shallow dishes or ice cube trays and partially freeze it. When the top begins to harden slightly, remove the cream from the dishes and whip it all together in a large bowl until its consistency is smooth and creamy, taking care that it does not melt completely. Pour the cream into waxed tubs or cartons, leaving expansion space, and return to the freezer for solid freezing.

Making ice-cream sticks is fun, too, and very popular with children. To do this, follow the above procedure but return the whipped cream to ice-cube trays instead of cartons. When they are frozen, pop them out of the trays carefully and insert pointed sticks made of wood or rolled paper (they can be bought in quantity from many chain stores). Next, dip the cubes-on-sticks in melted chocolate, crushed nuts or shredded cocoanut and slip each one into a small glassine sandwich bag. Stand them upright, sticks up, in a cardboard box and put them in a safe place in the freezer, where they will not be crushed by other packages.

During the summer months, I invariably have a generous supply of these treats on hand. If I have been too lazy to make them myself, I rush out into the street when I hear the gay tinkle of the Good Humor man's cart and buy several dozen in assorted flavors, keeping them in the manner described above. This does not save any money, but it makes me popular with my small friends.

16. Freezing Vegetables from A to Z

The home freezer is blessed with heartfelt fervor by housewives all over the land for its three major contributions to daily life—convenience, economy and better nutrition.

These three benefits are dramatically demonstrated when it comes to using the freezer for the preservation of vegetables. Freezing removes the drudgeries of canning, and provides a better product.

With a patch of garden or by purchasing bushels of your favorite vegetables during their harvest times you can save enough money each year to buy yourself a new wardrobe of dresses. More important, perhaps, is the fact that you will never again have to rely on seasonal crops or pay premium prices for out-of-season vegetables. A freezer in your kitchen or on the back porch means garden-freshness and goodness for your table all year long.

When you have a freezer, and especially if your ownership of this marvelous appliance is fairly recent, there is a temptation to freeze everything and anything in sight. As the harvest months for vegetables and fruits roll around, you may look into the chilly recesses of your freezer and see that you have quite a bit of unused space just waiting to be filled up with hundreds of packages of everything that grows on a bush, a tree or in the ground.

Before you go on a spree, however, remember that in another couple of months you will most probably be replenishing your supply of beef, which takes up quite a bit of freezer space. Take into careful consideration, too, your family's tastes and preferences for vegetables. It would be foolish and wasteful to freeze a bushel of lima beans just because they are cheap if your family can't be induced to eat lima beans more than once every two months.

About a year ago, while driving in the country, I passed a handsome farm fronted by a roadside stand on which were displayed hundreds of baskets brimming over with lovely produce. Needless to say, I stopped the car and bought very inexpensively several large baskets of a few vegetables, fruits and melons destined for my freezer. After the transaction was satisfactorily completed, the farmer waved his arm at a great bin holding more zucchini in more shapes and sizes than I had ever seen.

"Help yourself," he said. "This year's crop came in better than I expected. I'll never sell 'em all."

I picked out a dozen or so of cucumber size and started

back to the car, but the farmer came after me with one of those tall baskets filled to the spilling-over point.

"Here," he said. "You told me you've got a freezer. Well, freeze 'em!"

I froze 'em. I gave as many fresh ones to friends and neighbors as they would accept, and froze the rest. I spent rather a lot of time doing it, and ended up with 40 impressive quarts of zucchini—sliced, blanched and packed. During the months to come, I used zucchini as a steamed vegetable, in casseroles, cold with vinegar and oil, and french fried. Because it was in my freezer in such quantity I felt I had to use it.

I like zucchini, but I got awfully tired of it. The time arrived when my friends, invited to dinner, would say with what was meant to be casual interest, "By the way, have you used up that zucchini?"

Besides, the zucchini season coincides with the sweet corn harvest in my part of the country, and if my beautiful upright freezer had developed an unheard-of mechanical breakdown at that time, I could have kept it going at subzero temperatures with the frigid looks I gave those space-consuming packages of zucchini.

I can't stand waste. Throwing away good food, or letting it spoil, seems to me as immoral as larceny. When I was a little girl and rejected or disdained the food placed in front of me, my mother used to say, "Think of the starving children in India and be ashamed of yourself." I still think of them. I think of the large-eyed, skinny, belly-distended children I saw in the Indian villages of Mexico recently. I cannot throw food away.

In the height of the vegetable season, plan your freezing days with intelligence and foresight. If you grow vegetables in your own garden, learn the *exact moment* when the various crops should be harvested to capture their peak quality.

If you buy vegetables in quantity from neighboring farms or from markets which receive fresh shipments daily, get there early and run all the way home with your purchases, for sun, air and heat are racing against you to rob your vegetables of their valuable vitamin content.

Practically all garden vegetables can be frozen (except lettuce and a few similar fresh salad greens) although some of them take to freezing better than others do.

While the varieties recommended for freezing throughout this chapter have been determined by personal experience and poring over the information made available to me by nurseries, college experimental stations, commercial frozen food companies and the government, there is always the possibility that the area in which you live presents specific local problems of produce gardening which you should take into consideration before buying packets or pounds of seed, or investing in young plant shoots.

If, instead of growing them you plan to buy large quantities of vegetables for freezing and have had no previous experience, it would be wise to try out three or four packages in your freezer before going all out in your purchases. Do this by buying a few pounds of the vegetable you are in doubt about. Buy them very early in the season of the individual variety. Prepare them according to the recommendations here, freeze them, and sample them within a few days.

This test will not provide any information about the effects of long-term storage, but it will satisfy you as to whether or not freezing itself is suitable for the vegetable according to your family's tastes.

Preliminary Instructions for Freezing Vegetables

Before attempting to freeze large quantities of vegetables, check your kitchen equipment to make sure you have on hand everything you will need. You will want several sharp knives, a cutting board, at least an 8-quart-capacity pot with a strainer insert and a tight cover for water blanching, a rack or trivet for steam blanching, enough packaging material to contain all the vegetables you prepare, and plenty of ice for cooling the vegetables after they are blanched.

Is Blanching Necessary?

Yes.

Absolutely, and ninety-nine per cent without exception.

Every once in a while, a brightly worded article will appear in a magazine which says, in effect, "Ladies, you don't have to blanch vegetables for freezing. That's just an old fuddy-duddy idea. Just wash the vegetables, pop them into containers and toss them into the freezer. They'll taste and look every bit as good when you eat them, so why fuss?"

Here's why: Blanching, which is a form of partial cooking at high heat, stops enzymic action which, if allowed to continue, will surely change the color, flavor, texture and nutritive value of vegetables.

While it is true that *some* varieties of vegetables may possibly look and taste "every bit as good" if they are frozen without having been blanched, and are eaten within a few months, you might just as well be eating vegetable-flavored diced inner tubes for all the nourishment you'll get out of them. Their vitamins will have packed up and left forever.

You may say, "Oh, well, what's a vitamin more or less?" If that is your attitude, and you don't particularly mind eating vegetables whose flavor, color and texture are less than wonderful, then go ahead and freeze them without blanching. Your freezer will still be a monument to economy and convenience, if not to health and nutrition. Among other things, blanching removes dirt from the vegetables' pores, destroys from 90 to 100 per cent of bacteria which may be present and makes certain that those chemical sprays, in general use nowadays on most farms, are safely washed away.

Besides, in this era of over-processed, devitalized foods, every vitamin I can rescue for the well-being of my inner chemistry is highly prized. I am far from being a food faddist, but I like the food I eat to be useful as well as decorative and palatable. It's great fun to be healthy.

When you freeze kale, for example, you freeze about 20,000 units of Vitamin A and 600 micrograms of Vitamin B-2 (riboflavin) with every 3½ ounces of the curly green leaves. That is to say, you freeze approximately these considerable amounts of vital food elements if you have blanched the kale for 60 seconds. Vitamin A is essential to good eyesight, teeth, bones, skin and soft tissues; Vitamin B-2 is required for systemic well-being, for the nerves, and as a "burning" assistant in the utilization of starches and sugars. The 3½ ounces of kale, properly protected against vitamin loss, supply the body's daily requirements of both essential food elements by government standards.

Green peppers are frequently mentioned by some writers as vegetables which do not require blanching; but 2½ ounces of peppers, shielded against devitalization, yield almost 90 milligrams of Vitamin C, that most volatile of elements which must be taken in large quantities each day for proper functioning of the body.

What's a vitamin more or less? Medical science has proved that without sufficient intake of vitamins (and other important food elements, of course), we would literally starve to death even if our appetites were satisfied by huge helpings of foods which fill but do not feed.

Your freezer's valuable service as a safety-deposit box for vitamins, minerals and proteins should not be regarded lightly. For those who are interested in the nutritional components of the various foods they eat, I shall include a brief analysis of each vegetable's nutritive value along with the rest of the information germane to its selection, preparation, packaging and freezing.

How to Blanch Vegetables for Freezing

There are two recommended methods: 1, blanching in water; 2, blanching in steam. A few varieties may be treated in other ways: The oven may be used for heating pumpkin, potatoes and winter squash; mushrooms may, if desired, be sautéed lightly in butter; tomatoes for juice or sauces may be simmered gently.

WATER BLANCHING

Put the water up to boil before you start the rest of your preparations. The faster you can get vegetables into the freezer, the better they will be.

Use a pot big enough to hold at least two gallons of furiously boiling water. The pot should have a tight-fitting cover and should be the type into which you can fit a fine-mesh wire basket or strainer with handles. While you can use large squares of cheesecloth instead of a basket, this method is somewhat clumsy and, by confining the vegetables instead of allowing them to circulate freely in the boiling water, you may under-blanch some and over-blanch others.

There are special blanching pots available in most housewares departments, made of aluminum or enamel with a fitted colander-like insert. This is the type I use—not only for freezer blanching, but also for many other purposes such as cooking spaghetti and making soups which I strain.

Use aluminum, steel or enamel pots only, never iron oı copper.

It is generally best to blanch only about one pound (two cups) of vegetables at one time, although small vegetables such as peas or beans, which can be agitated in the water to distribute the heat evenly, may be increased to two pounds.

Allow at least one gallon of boiling water for each pound of vegetables.

Prepare vegetables as you would for table use, washing them thoroughly and removing over-ripe, damaged or immature ones. Sort according to size, for smaller vegetables take less blanching time than larger ones of the same variety.

Put vegetables into the pot containing the basket (which should, for best results, be allowed to remain in the boiling water except when it is necessary to raise the blanched vegetables). If your water is boiling hard and you do not put in more than one pound of vegetables per gallon of water, the boil should not be disturbed for more than a few seconds. Keep the heat high and cover the pot tightly. The purpose is to raise the temperature of the vegetables to 212°F. for the number of minutes recommended for each variety. At sea-level, water boils at 212°F.

If you have followed the pound-per-gallon measure and have kept the heat high, start counting the blanching time as soon as you cover the pot. If you are in doubt about the weight of the vegetables, or your blanching pot is smaller than a two-gallon one, or the water seems to have quieted down after you've added the vegetables, start the time count from the instant you see boil-bubbles.

If I have made this procedure seem pretty rigid, it is because I am assuming that you want your frozen vegetables to be as good as you can get them. Nothing very terrible will happen if you err on one side or the other of blanching time. Too little blanching will bring about some color change and vitamin loss in the frozen product; too much blanching will sacrifice some crispness and the garden-fresh flavor you can expect if you manage to hit the timing on the nose.

[IMPORTANT: *If you live 5,000 feet or more above sea-level, it will be necessary to add one minute to the blanching times given for each vegetable.*]

When the blanching time is reached, remove the vegetables immediately from the boiling water and *cool thoroughly,* allowing at least the same amount of time for cooling as you did for blanching.

COOLING

It is necessary to chill vegetables well before packaging them, otherwise they will continue to cook in their own heat. Cooling may be done by plunging the vegetables immediately into a large quantity of very cold water—at least 50°F. Change the water frequently to maintain this temperature. If your tap water runs this cold, you may hold the vegetables under a spray until they are cooled. I prefer to use ice water.

You can, as I do, freeze large blocks and cylinders of water in bread tins and saucepans the night before your vegetable session. Another way is to order a 25-pound cake of ice from your iceman. I probably would, to conserve the electric power it takes to freeze water, if I didn't live in a neighborhood which hasn't seen an "ICE" card hung in the window for years.

You will find that cooling frequently presents problems you had not foreseen. The day you process a bushel of vegetables may be a very hot day, and the ice you thought would be ample just melts away, maddeningly, as you introduce the 212°F. vegetables a pound at a time. At the tail end of your bushel you may find yourself with tepid "ice-water" and not enough cubes in the refrigerator to do much good.

Quite by accident, I recently discovered what has turned out to be a perfect solution of the cooling problem.

I am addicted to picnics and picnic fixings while driving around the country, and for some years have possessed not only a fitted wicker basket but also a wonderful insulated picnic pail called, with typical advertising poetry, the "Pik-Nik Pail." This is a four-gallon cylindrical container whose superb fiberglass insulation and several other mysteriously effective insulating properties keep cold things cold or hot things hot for hours and hours.

I have used my handsome blue "Pik-Nik Pail" with its cheerful yellow lid for everything from keeping boiled lobsters hot while driving in Maine to keeping bottled water or beer drinkable while driving in Mexico. I know how good its insulating ability is, for once I filled it with ice cubes for a New Year's Eve party and found solidly frozen cubes in it two days later.

Not long ago, I was in the middle of preparing a bushel of green peppers for the freezer. It was a hot day, and I was worried about the staying powers of the two 6-quart saucepans of ice I had frozen for the chilling process. I had planned to do the cooling in a large tub, but discovered that it was temporarily in use elsewhere. I cannot use my single sink for cooling because I need it for washing each batch of vegetables before blanching them.

Stumbling around in the cellar looking for another cooling vessel, I fell over the "Pik-Nik Pail" and eyed it speculatively. The pail's mouth is 10 inches wide, and I found that I could get my ice cakes into it with ease. I covered the ice with its own volume of water, and then discovered what any high-school physics student knows. The ice floated.

This meant that my peppers would undoubtedly swim

underneath the ice and be a nuisance to chase with a sieve.

I reached for one of the prizes of my gadget collection—a collapsible French basket made of fine wire mesh which is used for washing and draining salad greens. I'm shot with luck, apparently, because the basket fitted perfectly into the pail, forcing the ice below the surface of the water.

I proceeded happily with washing, weighing and blanching the peppers and poured them into the salad basket which, in turn, was inserted into the "Pik-Nik Pail" and allowed to remain there for the proper chilling time—which is a little more than the time it takes to blanch any given vegetable.

I was able to lift the basket out of the pail, drain the vegetables in record time and proceed with packaging, sealing and freezing. The wonderful part of it all is that the "Pik-Nik Pail" maintained the iciness of the water until the entire bushel was processed. Just for fun, I left the water in the pail until the next day, and it was still ice cold!

This is the best method I have ever found for chilling blanched vegetables. You can probably find a "Pik-Nik Pail" in your favorite hardware store or housewares department, or a flossier plaid-printed version called the "Skotch Kooler." Either one is a good investment, not only to make freezing easier, but also for its original function as a thermos container for hot or cold foods.

When the vegetables are cool, drain them thoroughly in a colander and then turn them out onto absorbent towels, either cloth or paper.

Pack at once according to the instructions for each variety.

STEAM BLANCHING

This may be done in your pressure cooker, using a rack or trivet with legs long enough to clear completely two inches of water in the bottom of the pan. Steam blanching may also be done in an ordinary kettle with a tight lid and a rack which holds a steaming basket at least three inches above the bottom of the kettle.

Prepare the vegetables as you would for table use and put them in the basket only one layer deep at a time. Bring the water to boil, lower the basket onto the steaming rack and close the lid tightly. *Note:* If you use a pressure cooker for steam blanching, do not close the petcock and do not clamp the top down. No steam pressure is needed. Start counting the steaming time as soon as the lid is on.

[IMPORTANT: *If you live in an area which is 5,000 feet or more above sea-level, add one minute to the time specified for each vegetable.*]

When the steaming time is finished, cool the vegetables immediately, using the same methods recommended for water-blanched vegetables.

Oven Heating

If you heat pumpkin, squash, sweet potatoes or yams for freezing, place a pan of water in the oven along with the food to prevent drying out. Pre-heat the oven to 450°-500°F. before putting the food in.

Which Method of Blanching Should You Use?

Steaming is preferred by many home economists for certain varieties of vegetables, especially those which are cut or sliced, for this method preserves more of the minerals and vegetables than water does.

Leafy vegetables such as kale, spinach and other greens

should never be steamed, however, for they have a tendency to mat together.

Despite my jealous regard for vital food elements, I ordinarily prefer water blanching. It is easier, and I am surer of my product than I am when I use steam. Besides, I am very likely to save some of the water in which I have blanched such vegetables as artichokes, asparagus and leafy greens, combining them and using the liquid when I make soup stocks and gravies from bones and meat trimmings. It is very seldom that anything goes to waste around here!

Freezing Vegetables

Artichokes

These are not supposed to be freezable, but I have frozen them with enough success to recommend them to other artichoke-lovers. Once I was lucky enough to live on the West Coast and never ceased to wonder that I could actually go outside in my own back yard and pick as many artichokes as I thought I could eat for dinner. At that time, unfortunately, I had no freezer. Nowadays, when the artichokes are green and inviting in my neighborhood market, I select several dozen for short-term freezer storage, seldom more than a few months.

Best varieties for freezing: The only ones I know about are those grown in the coastal area of California. Select small, uniformly green, compact artichokes whose leaves are not yet open. They should look like buds, and the tips should be innocent of the sharp spikes which come with maturity.

Nutritive value: Good source of Vitamins A, B-1 and C. Fair source of calcium and phosphorus.

Preparing for the freezer: My method for doing this may be a little unorthodox, but it is the way I have discovered to be most satisfactory. First, I remove the outermost leaves and discard them. Next, I trim the stalk to about one inch, and with a very sharp knife cut across the top of the 'choke to expose the layers of tightly packed inner leaves. I blanch six at a time in two quarts of boiling water to which I have added the juice of one lemon, some celery stalks and leaves, a peeled clove of garlic and half a bay leaf.

Blanching time—three minutes for small artichokes, four for medium-sized ones; I never freeze the very large ones.

After cooling them thoroughly in ice water, I drain them upside down on paper towels. I wrap each one closely in cellophane, sealing the open flaps with freezer tape, and put six of them in a chicken-size polyethylene bag, making a goose-neck closure.

I honestly don't know how long artichokes can remain in the freezer, for I usually serve mine before four months' time has elapsed.

Asparagus

This is one of the best freezing vegetables I know, and I am especially glad about it because I live in a part of the country where the asparagus season is short. Around here, asparagus is plentiful and cheap during May and June, becoming scarcer and much more expensive later on in the year.

If you have a garden, you can grow either white or green asparagus, for the color depends only on how it is grown. White asparagus is grown entirely underground and is harvested as soon as the tips thrust through the surface of the soil. Green asparagus is grown above the ground and gets its color from exposure to sunlight.

Best varieties for freezing: Giant Washington, Mary Washington, Paradise.

Nutritive value: Good source of Vitamins A and C, also phosphorus and iron.

Approximate yield:

One pound of uniform stalks equals one pint container.

One crate (24 pounds) equals between 16 and 20 pints.

Preparing for the freezer: First, remove the bottom-most woody portions and discard them. Next, bunch the asparagus together so that all the tops are even, and cut into five- or six-inch stalks. The ends which remain can be blanched separately and used for salads or in soup. Wash thoroughly in running water. I usually remove the little scales that appear on the stalks, for these may harbor sand. Sort the spears into three groups of even sizes—small, medium and large. Do this by comparing the butt ends.

Blanching in water:

Large spears, 4 minutes.

Medium spears, 3 minutes.

Small spears, 2 minutes.

Blanching in steam: Tie uniform sizes of asparagus lightly with white string and place them heads-up on the steaming rack. Add one minute to the times given for water blanching.

Cool immediately and drain thoroughly.

Packaging: After the asparagus is cooled and drained, I mix up the sizes again, dividing each quantity to be frozen into equal numbers of small, medium and large spears. For packaging, I use whatever I happen to have the most of at the moment—trunk-opening boxes lined or outer-wrapped with cellophane; tall plastic boxes, or even laminated bags.

How many you put in each package should depend on the number of persons in your family. If you have a large family and serve asparagus generously, a good package can be made by alternating the heads of the asparagus and placing a large quantity in the middle of a sheet of cellophane or laminated freezer paper. Fold the paper over the asparagus from all sides, then roll it into as small a bundle as you can, sealing the flap.

Beans

There are so many varieties of beans that each of them should be regarded as an entirely different vegetable, for the methods of preparing and blanching them will vary according to type.

Lima Beans

Best varieties for freezing:

Baby limas: Henderson or Burpee bush, baby Fordhook.

Large limas: Fordhook, Challenger, King of the Garden.

Harvest or purchase lima beans when only about 5 to 10 per cent of the beans have changed from green to white. White lima beans are indicative of a high starch content.

Nutritive value: Good protein source, but somewhat high in carbohydrate content. Good source of Vitamins B-1 and C.

Approximate yield:

Two pounds of limas in the pod equal one pint.

In bulk, one bushel (32 pounds) will yield from 12 to 16 pints, depending on the size of the beans and the amount of waste.

Preparing for the freezer: Shell and sort according to size. Discard all withered, discolored or mealy beans.

Blanching in water:

Large beans, 4 minutes.

Medium beans, 3 minutes.

Small beans, 2 minutes.

Blanching in steam: Add one minute to each of the above times. Cool and drain.

Packaging: Lima beans (indeed, most vegetables) may be packed in laminated or polyethylene bags, in bag-and-box combinations, or in plastic containers. If you use rigid containers, allow about ½-inch expansion room at the top.

Snap, Green or Wax Beans

Best varieties for freezing:

Bush: Tendergreen, Giant Stringless, Green Pod, Full Measure.

Pole: Kentucky Wonder, Blue Lake, White Creaseback.

In the summer, harvest or purchase before there is much development of beans within the pods, for the fibrous material of the pods increases with maturity. Fall crops may run to larger pods. Both varieties may be frozen, although you will get a better product, usually, if the pods are under 5 inches long.

Nutritive value: Good source of Vitamins A, B-1 and C.

Approximate yield:

One pound (or less) equals one pint.

One bushel (30 pounds) equals from 30 to 45 pints.

Preparing for the freezer: Wash thoroughly, remove stem ends (and strings, if any). Leave whole, or cut into pieces one or two inches long; or, slice lengthwise into strips for french-style (julienne) beans.

Blanching in water:

Whole beans, 3 minutes.

Cut beans, 2 minutes.

Frenched beans, 1 minute.

Blanching in steam: Add one minute to each of the above times. Fast and thorough cooling is absolutely essential. Use quantities of ice if necessary. Drain on absorbent towels.

Packaging: Pack in rigid containers, leaving ½-inch expansion space; in laminated or polyethylene bags, or in bag-lined boxes.

Soybeans

This fine protein food is, happily, excellent for freezing.

Best varieties for freezing: Jogun, Bansei, Giant Green, Easy Cook.

Nutritive value: High in vegetable protein and in B vitamins.

Approximate yield:

2 to 2½ pounds (in pod) equal 1 pint.

1 bushel (30 pounds) equals from 12 to 15 pints.

Preparing for the freezer: Harvest or purchase while the pods are firm, plump and bright green. Wash pods under running water. Put pods in vigorously boiling water and cook for five minutes. Cool rapidly in cold or ice water.

Squeeze the beans out of their pods and pack immediately in rigid containers, leaving ½-inch head space, or in laminated or polyethylene bags. No further blanching is necessary for soybeans prepared in this manner.

Beets

Unless you are powerfully partial to this vegetable, or unless your family consumes quantities of beet soup (borscht) at all seasons of the year, you may decide to skip this one for the freezer, preferring to preserve it by canning. It is entirely possible to freeze beets, either whole or diced or sliced, but they are not especially superior to canned varieties for the reason that they must be frozen *pre-cooked,* not merely blanched, for best results. Short blanching and long storage tend to toughen beets and make them rubbery.

Because I like to serve beets with beef à la Stroganoff, I usually freeze the very small, tender varieties whole and later prepare them for the table with a cooked sauce composed chiefly of sour cream, dry mustard and horseradish.

For sliced sweet-and-sour beets, I invariably use the excellent canned varieties sold by leading chain stores.

Best varieties for freezing: Detroit Dark Red, Early Wonder, Crosby.

Nutritive value: Fair source of minerals and Vitamins A, B-1, B-2 and C.

Approximate yield (without tops):

1 to 1½ pounds equal one pint.

1 bushel (52 pounds) equals from 35 to 45 pints.

Preparing for the freezer: Wash thoroughly and sort according to size. Trim tops closely, and *save the greens.* Do not slice or cut beets at this time.

Cooking (not blanching): Cook in water until tender, ready for eating. Depending on the size of the beets, this will be anywhere between 20 and 45 minutes. If the beets you are freezing are of various sizes, put the largest ones in the boiling water first, adding the others of graduated sizes at ten-minute intervals.

When beets are tender, cool them promptly under running water. Peel or rub off skins of mature beets before slicing or dicing them. Whole young beets of small size, no larger than one inch in diameter, may be frozen without peeling.

Packaging: Pack cut beets in rigid containers, leaving ½-inch expansion space, or in bag-lined boxes. Whole beets may be packed in laminated or polyethyene bags.

Beet Greens

Nutritionally speaking, these are the best parts of beets. It is as though the roots merely act as a transient warehouse for the life-building food elements delivered by the good earth, holding just as many as needed for growth and color and sending the rest on up above ground in the form of vitamins and minerals.

I have seen a number of American housewives do an inexplicable thing in terms of nutrition and economy. When planning a meal, they frequently buy beets *and spinach* as vegetables. Spinach is fine, but the same amount or more nutritive value exists in the beet tops they cut off and discard! The taste is somewhat similar, too.

Nutritive value: Excellent source of calcium and iron; good source of Vitamins A and B-2.

Approximate yield:

1 to 1½ pounds equal one pint.

15 pounds equal from 10 to 15 pints.

Preparing for the freezer: Select the youngest, most tender leaves, removing tough stems and imperfect parts. Wash thoroughly, both under running water and in a tub bath. Cut with scissors into small pieces, or leave whole for chopping after blanching.

Blanching: In boiling water only, for 2 minutes.

Cool immediately in very cold running water or in ice water. Drain as completely as possible on paper or cloth towels.

Packaging: Pack leaves in polyethylene or laminated bags. Pack chopped greens in rigid containers, leaving ½-inch expansion space.

Broccoli

Select for freezing only young, tender shoots whose stalks are not more than one inch thick and whose heads are uniformly green.

Best varieties for freezing: Early Sprouting, Italian Sprouting, Riviera.

Nutritive value: Good source of calcium, phosphorus, Vitamins A and B-2.

Approximate yield:

One pound equals one pint.

One crate (about 25 pounds) equals from 20 to 24 pints.

Preparing for the freezer: Trim off large outer leaves and cut off tough, woody parts of stems. Wash thoroughly. If you suspect the presence of garden insects, soak the stalks in a salt solution for half an hour, using 4 teaspoonfuls of salt to each gallon of water used. Wash again to remove saltiness.

Cut lengthwise into pieces as uniform as possible, leaving the flower-like heads about 1½ inches across. This will allow the broccoli to blanch evenly and will make it easier to pack as well as more attractive to serve.

Blanching in water:

For large heads, 4 minutes.

For 1½-inch heads, 3 minutes.

For smaller heads, 2 minutes.

Blanching in steam: Add one minute to each of the above times.

Cool by plunging the blanching basket into very cold water. Drain as completely as possible on absorbent cloth or paper towels.

Packaging: Using kitchen tongs, carefully lift stalks and pack them loosely in long polyethylene or laminated bags, or in rigid plastic containers, placing the heads alternately in opposite directions.

Brussels Sprouts

Select sprouts whose color is rather dark green. The little cabbage-like heads should be firm and compact and free of insects.

Best varieties for freezing: Hall Dwarf, Long Island Improved.

Nutritive value: Good source of phosphorus, iron and Vitamin B-1.

Approximate yield:

One pound equals one pint.

Eight quart boxes equal 12 pints.

Preparing for the freezer: Cut off the stems close to the heads and trim off any tough or wilted outer leaves. Sort the sprouts according to size—small, medium and large. The largest ones should not be more than two inches in diameter. Wash thoroughly. If sprouts show evidence of the presence of garden insects, follow the salt-water soaking instructions given for Broccoli.

Blanching in water:

Large heads, 5 minutes.

Medium heads, 4 minutes.

Small heads, 3 minutes.

Blanching in steam: Add one minute to each of the above times.

Cool under running water or in ice water. Drain almost dry.

Packaging: Pack closely in laminated or polyethylene bags, or in rigid containers. It is not necessary to leave expansion space for Brussels sprouts, provided you have drained them well.

Cabbage

Because this vegetable is usually cheap and plentiful during most seasons of the year, it is rarely frozen as an economy measure. It is, however, a good vegetable to have on hand to be served up with ham or corned beef, and can be added to soups or stews for extra flavor and food value. Catch your cabbage for freezing while it is young, either from your own garden or during the early part of each variety's peak season.

Best varieties for freezing: Copenhagen, Hollander, Danish Red.

Nutritive value: Excellent source of calcium and Vitamin C.

Approximate yield: One small head equals one pint.

Preparing for the freezer: Trim the coarse outer leaves from the heads. Depending on how your family likes this vegetable, separate the head into leaves, grate it coarsely, or cut it into fairly thin wedges. Wash thoroughly.

Blanching in water:

Wedges, 3 minutes.

Leaves, 2 minutes.

Shreds, 1½ minutes.

Blanching in steam: Add one minute to each of the above times.

Cool thoroughly under running water or in ice water, and drain well.

Packaging: Pack shreds or leaves in laminated or polyethylene bags, wedges in rigid containers, leaving ½-inch head space.

Carrots

Select for freezing only young, tender carrots, preferably small ones.

Best varieties for freezing: Nantes Half-long or Coreless, Danvers, Imperator, Chantenay.

Nutritive value: Superb source of Vitamin A (carotene) if not over-cooked; good source of Vitamins B-1 and B-2; fair source of calcium and iron.

Approximate yield:

1 or 1½ pounds equal 1 pint.

1 bushel (50 pounds) equals from 32 to 40 pints.

Preparing for the freezer: Leave the smallest carrots whole, washing them carefully but not scraping them, in order to preserve the maximum amount of nutrients. Cut larger carrots, after scraping, into finger-length halves or quarters. They may also be cut into ½-inch round slices, or diced into cubes no larger than ½ inch.

Blanching in water:

Whole carrots, 3½ minutes.

Fingers, 3 minutes.

Sliced or diced, 2 minutes.

Blanching in steam: Add 2 minutes to each of the above times.

Cool quickly in ice water, and drain well.

Packaging: Pack whole carrots or fingers in rigid containers, leaving ½-inch expansion space. Pack slices or cubes in laminated or polyethylene bags.

NOTE: Carrots and peas may be packaged and frozen together, preferably in rigid containers, but each vegetable should be prepared and blanched separately according to instructions.

Cauliflower

This vegetable, like asparagus and broccoli, is one of the best freezers of all, although a certain amount of extra care must be taken to safeguard the pretty white flowrets against discoloration. Inasmuch as the bleaching agent is added to the water in which cauliflower is blanched, I do not recommend steam-blanching except for the purple varieties, which of course do not require bleaching.

Best varieties for freezing: Forbes, Snowball, Perfection, Erfurt, White Mountain, Purple.

Nutritive value: Good source of calcium, Vitamins B-1 and B-2.

Approximate yield:

One pound equals one pint.

Six medium-sized heads equal about nine pints.

Preparing for the freezer: Cut stem close to the head. Break or cut into flowrets and pieces about one inch across.

Wash, removing any insects which may be present by soaking pieces for half an hour in a salt-water solution (4 teaspoonfuls of salt to each gallon of water used). Wash again to remove saltiness, and drain as completely as possible.

Blanching in water: Add to one gallon of boiling water 2 teaspoonfuls of powdered citric acid, or the juice of one lemon, or follow manufacturer's directions on package of citric-ascorbic acid powder used for fruits. Blanch for 3 minutes in this solution if pieces are no larger than 1 inch; blanch larger pieces for 4 minutes.

Blanching in steam (purple variety only):

Large pieces, 5 minutes.

Small pieces, 4 minutes.

Packaging: Drain well and pack in plastic bags, or in bag-and-box combinations.

Celery

The time may yet come when this crisp, crunchy vegetable may be frozen in raw stalks, for salad and appetizer use, but at present it may only be frozen for use as a cooked vegetable. Because it does make a delicious side dish when served either plain or with a cream-base cheese sauce, and because it adds so much flavor to soups, stews and casseroles, I often freeze several containers of celery during the seasons of the year when the price is lowest. I prefer the green pascal type to the bleached variety.

Nutritive value: Cooking is likely to dissipate this delicate vegetable's vitamin content, although iron and most of its calcium are retained; but it is so extremely low in calories that it is a virtuous vegetable for reducing diets—if served plain, of course.

Approximate yield: Two medium-sized bunches equal one pint.

Preparing for the freezer: Use only crisp, tender stalks. Remove strings, if necessary Scrub stalks well under running water, removing all grit and dirt. Cut into 1-inch lengths.

Blanching in water: 3 minutes.

Blanching in steam: 4 minutes.

Cool immediately, preferably in water to which you have added a considerable quantity of ice.

Packaging: Drain and pack dry in laminated or polyethylene bags. If you like, you may add enough of the blanching water to cover celery placed in rigid containers, leaving ½-inch expansion space. This latter method is especially good for celery which is to be added to soups, for the liquid can be added also.

Collard Greens

Select young, tender leaves and prepare them according to the instructions for Beet Greens. Blanch in water only for 3 minutes, never in steam.

Corn on the Cob

There is so much controversy among freezer devotees about the correct method for freezing corn on the cob that I can almost anticipate the number of letters I will receive from those who belong to the "freeze-'em-in-the-husk" and "blanch-'em-in-milk" schools.

Next to artichokes and running a close second to mushrooms, sweet corn on the cob is my favorite vegetable. You may be sure I have tried every known method of freezing it, and a few which do not exist. I am even perfectly willing to give up the tremendous amount of freezer space required for the cobs, frequently finding that Experiment Number 99 was a complete failure—two cubic feet of failure, at that!

As far as I am concerned, the best corn in the world came from my own mulch garden. I planted only one variety, a special seed developed by the Joseph Harris Company of Rochester, New York. It's called "Wonderful" and it is aptly named. This was when I lived in a Connecticut farm house, some years ago. Every August I shared my abundant harvest with a group of friends who learned to anticipate the annual corn party. My cornfield was close by the outdoor barbecue. On the afternoon of the feast, a hot fire was allowed to burn itself out into a deep bed of winking embers. Halfway between the cornfield and the barbecue we placed a huge tub of salted spring water. As the sun sank low in the sky, we stripped young ears of corn from their stalks, removed the silks, raced to the tub of water and swished them around for a while, then ran to the barbecue and tossed them onto the coals. They steamed and sizzled, and then they were done.

We husked them, not altogether reckless of scorched fingers, but almost. Buttered and salted, the kernels seemed to leap from the cobs.

Corn has never tasted quite so good since, for I have been long gone from my Connecticut cornfield. Even local farmers' produce brought by pick-up truck to my Cape Cod door—and Cape Cod corn is deservedly popular—seems to lack the quality of those harvests. It has occurred to me that the quality may be less in the corn or in geography than in nostalgic memory.

When I started using a home freezer, one of the visions which danced through my head was that of serving steaming platters piled high with golden ears of corn to January dinner guests. This I did, but I must admit that on occasion my guests used to eye me pityingly, no doubt thinking my palate was so easily satisfied that I regarded this poor excuse for corn as delicious. There was little sense to apologizing or explaining that the current batch was from Experiment 37. The results spoke for themselves, and were not complimentary to my culinary reputation.

In more recent years, however, I have found what to me, at least, is the most rewarding method of freezing corn on the cob. I must say quickly, however, that for this vegetable it is not only correct freezing which is important; the cooking procedures are equally vital. (See page 327.)

Best varieties for freezing: Golden Bantam, Golden Cross, Wonderful. (Try if you must the evergreen, country gentleman and Mexican varieties. I have never found them successful.)

Nutritive value: Not exceptional, except for fair amounts of Vitamin B-1, and bulging with calories, but who cares?

Preparing for the freezer: Live close to a cornfield. If you can't rearrange your housing, select *guaranteed* young, fresh-picked ears whose husks hug the cobs in a tight embrace. Never buy husked corn or corn whose silk has been removed. Do not strip a section of the husk away from the ear to look at the kernels until you are ready for blanching and freezing. At that time, give an ear of a uniform crop the thumb-nail test. If the milk spurts, the corn is prime for freezing.

Husk corn quickly, removing all silk, and wash the ears under running water. Discard all starchy (over-ripe) ears and those with shrunken or under-sized kernels. I don't mean that you should throw them away. Put them aside for whole kernel or cream-style packaging. Sort ears according to length and diameter.

Blanching in water (use quantities of rapidly boiling water):

Large ears, 6-7 minutes.
Medium ears, 5 minutes.
Small ears, 3-4 minutes.

Blanching in steam: (I never do.) Add one minute to the above times.

Cool immediately and quickly in icy ice water. Wipe each ear dry with a clean towel or cloth.

Packaging: Wrap each ear individually in cellophane or

plastic freezer film. Freeze immediately in the coldest part of your freezer. If you wish, wrapped and frozen ears may be gathered up and put into large polyethylene bags for easier storage and accessibility.

Alternate method: A somewhat extravagant but effective way to handle corn on the cob for freezing is worth a try if you have not been satisfied with other results. Remove husks and wrap each ear individually in heavyweight aluminum foil. Place the closely wrapped ears in rapidly boiling water and add one minute to the above blanching times. Plunge corn immediately into quantities of ice-cold water (use plenty of ice) and chill for a period of time that is *double the blanching time*—12 minutes for small ears, 16 for medium, 22 for large. Wipe the foil-wrapped ears dry and freeze at once.

To serve corn packaged in this way, drop it directly from the freezer into cold water and bring it to the boiling point, then remove it, unwrap it and serve at once.

Special note for upright freezer owners: If for some reason you want to freeze corn on the cob for a very short period of time, no longer than two months, you may freeze them without blanching in husks which have not been disturbed in any way. Bend the silk back onto the husks and tie them tightly with a string, or slip a rubber band around them to hold them in place. Put corn in direct contact with a freezer coil, or on the coldest freezer plate shelf. After the ears are solidly frozen, package several together in polyethylene bags or wrap them in laminated freezer paper.

Thaw corn frozen in this way in cold water before husking the ears. Cook until tender in boiling water, or roast thawed ears still in their husks in a moderately hot oven, being sure to place a pan with about ½ inch of water in its bottom on another shelf of the oven.

Corn, Cream Style

Country gentleman may be used successfully for this method, but I still prefer the yellow varieties.

Follow the instructions for corn on the cob until you get to the blanching part.

Blanch corn in water for 4 minutes, unless the ears are exceptionally large, in which case they should be blanched for 5 minutes. Cool rapidly in ice water. Using a razor-sharp knife, cut corn from the cobs about halfway through their kernels. Put all of these half-kernels in a bowl and then, using the back of a knife, scrape the cobs over the bowl to catch the hearts of the kernels and the milky juice. Stir well, and put the bowl of corn in your refrigerator, covered, to chill. When thoroughly cold, pack the cream-style corn in rigid containers, leaving ½-inch expansion space. Freeze quickly.

Corn, Whole Kernels

Proceed as if for cream-style corn, but cut the kernels as close to the cobs as possible without including any of the tough part of the cobs. Package in laminated or polyethylene bags, or in rigid containers.

Cucumbers

This is another of the supposedly unfreezable vegetables which I freeze. Not, to be sure, in their natural state. I slice peeled cucumbers directly into plastic boxes containing a mixture of half water, half white vinegar, to which I have added one teaspoon of sugar and ½ teaspoon of black pepper for each pint of liquid, filling the containers to within 1 inch of the top. Before I serve these, I defrost them slowly in the refrigerator. When they are completely thawed, I add onion slices and salt, and let them stand for several hours. *I* like them.

Eggplant

This is another favorite in my house, especially prepared *parmigiano* or french-fried in batter and crumbs. Unless you are sure to select firm, medium mature eggplant with tender seeds, the frozen product may be rubbery in texture.

Best varieties for freezing: Black Beauty, regional hybrids.

Preparing for the freezer:

Method 1. Peel washed eggplant and cut across in ⅓-inch slices. Do not peel more than one eggplant at a time to minimize discoloration.

Blanch in boiling water to which you have added 2 teaspoonfuls of ascorbic-citric acid powder for each gallon of water. (See page 275 for discussion of ascorbic-citric powder solutions.) Blanch for 4 minutes. Cool in ice water, drain completely on absorbent towels. Package by slip-sheeting each slice with double thicknesses of cellophane, then wrapping enough eggplant for a single meal in aluminum foil or in laminated freezer paper.

If you prefer steam blanching, blanch for 5 minutes, and add the ascorbic-citric acid powder to the cooling water.

Method 2. Peel, then either slice as above or cut into ½-inch sticks. Dip in a thin fritter batter, or in beaten egg yolks and then in bread crumbs. Fry in deep fat until light golden brown. Drain on absorbent paper towels or on brown paper bags and allow to cool. Package by slip-sheeting each slice with double thicknesses of cellophane, then wrapping enough eggplant for a single meal in aluminum foil or in laminated freezer paper.

Eggplant prepared and frozen by method 2 may be heated in a slow oven for serving or, when half-thawed, may be deep-fat fried to a darker color. Round or half-round slices may be used after thawing for *parmigiano*.

Kale

Select for freezing tightly curled kale which has been picked while the leaves are young and tender.

Best varieties for freezing: Dwarf Curled, Blue Scotch.

Nutritive value: Good source of iron, calcium, phosphorus, Vitamins B-1, B-2 and C.

Approximate yield:

1 pound equals 1 pint.

1 bushel (18 pounds) equals 12 to 18 pints.

Preparing for the freezer: Work with a small amount at a time, keeping the remainder to be frozen covered and under refrigeration in order to preserve the high nutritive content of this green vegetable. It is a good idea to put bunches of unwashed kale into the very large polyethylene bags and then in your refrigerator's hydrator or crisper while it is waiting to be cleaned and blanched. Discard any leaves which are tough, dried or yellowish.

Wash thoroughly under running water, and then wash again. Pull leaves from stems. If chopped kale is to be your goal, do not chop it at this time. Blanch the leaves in pieces, and chop them after the kale is cooled.

Blanch in water for 1 minute. Steaming is not recommended.

Cool thoroughly under running cold water, or in ice water. Drain completely and package with or without chopping.

Packaging: Pack leaves tightly in laminated or polyethylene bags; chopped kale in rigid containers, leaving ½-inch expansion space.

Kohlrabi

This rather homely but well-meaning vegetable is seldom found in sophisticated urban markets, but its subtle taste, combining the flavors of cabbage and turnip, makes it a favorite for creaming and for including in stews and soups.

Best variety for freezing: Early White Vienna.

Nutritive value: Not very high in any of the essential elements except calcium. However, its fibrous texture makes it a good bulkage food.

Preparing for the freezer: Select young, tender, mild-flavored kohlrabi which are not too large. Cut off tops and roots. Wash and peel. Leave small ones whole, dice larger ones in ½-inch cubes.

Blanching in water:

Whole, 3 minutes.

Diced, 1 minute.

Blanching in steam: Add one minute to above times.

Cool in cold or ice water, and drain on absorbent towels.

Packaging: Pack whole kohlrabi in laminated or polyethylene bags, diced in rigid containers, leaving ½-inch expansion space.

Kohlrabi Greens

Follow instructions for Beet Greens.

Mushrooms

Among freezer experts, mushrooms have been known to cause as many heated arguments as partisan politics. There is one faction which insists that the only way to freeze mushrooms is to pop them into containers without even so much as dusting them off, and freeze them immediately. Another clan believes in thorough peeling and steam-blanching. Still another group votes for water-blanching of tiny little buttons only, ignoring all other sizes or parts thereof.

Well, I am frank to admit my own partisanship as regards mushrooms, which to my mind are among the tastiest and most desirable of all vegetables. I never cease to marvel at Nature's benevolence in making even fungi delicious. This is another of the foods which do not provide a great deal of nutritive value, but whose sins of omission are forgiven because of epicurean contributions.

Because I am so partial to mushrooms, and because there are certain seasons of the year when they are prohibitively expensive, I go prowling for them during their "in" seasons and have worked out my own systems for stocking my larder against leaner times.

My system is simple, really. I merely keep my ultimate recipes in mind and prepare three or four different kinds of mushrooms. I like broiled mushrooms, the enormous ones, as topping for thick steaks or served on toast for lunch or late supper. So—I broil enormous washed but unpeeled mushrooms dotted with butter until they are not quite done, allow them to cool, and package them, cups-down, in plastic boxes.

I like little button mushrooms in recipes like beef Stroganoff, and so I sauté these, washed but unpeeled, in butter until *they* are not quite done. These I package in laminated bags, enough per bag for a recipe serving four.

I like sliced mushrooms sautéed in butter and garlic as a vegetable with Italian dishes, and so I slice peeled mushrooms into butter and allow them to sauté until they are almost done. I package them in rigid containers without adding the garlic, which I later sprinkle liberally from a garlic press over the thawed mushrooms as they are being heated.

I like bleached, smooth mushrooms to add to Chinese recipes. For these, I select little buttons and slices of larger buttons, blanching them in a small quantity of boiling water to which I have added the juice of one or more lemons, depending on how many I am doing at the time. (I have been known to freeze ten pounds of variously prepared mushrooms in one afternoon.) I cook them until they are almost ready to eat and freeze them in rigid containers with enough of the blanching water to cover them.

I like to have on hand plenty of mushroom stock for gravies, soups, stews and for use as a base for à la king recipes. Whenever, therefore, I prepare mushrooms of any size for freezing, I carefully put aside the stems and any peelings and cook them slowly for a long time in water to which I have added celery leaves.

Finally, breaking my own rule about freezing vegetables raw, I frequently do freeze small whole raw mushrooms—trimmed stems and all—in heavy plastic bags. These I cook, whole and while still frozen, or add to soups and stews.

Mustard Greens

Follow exactly all the instructions for preparing and freezing Kale.

Okra

Long given due respect as a basis for thick soup (its other name is gumbo) and stews, this lowly pod is coming into its own as a table vegetable of gustatory stature.

Best varieties for freezing: Clemson Spineless, Dwarf Prolific, Perkins Spineless, White Lightning.

Nutritive value: High in calcium, phosphorus, Vitamin B-1.

Approximate yield: 1 pound equals 1 pint.

Preparing for the freezer: Select young, tender pods. Remove stems, but do not cut into seedy parts. Sort according to size. Do not slice before blanching.

Blanching in water:

Large pods, 3 minutes.
Medium pods, 2 minutes.
Small pods, 1 minute.

Blanching in steam: Add one minute to the above times.

Cool thoroughly in ice water, and drain on absorbent towels. Slice, if desired, at this time.

Packaging: Pack whole pods in laminated or polyethylene bags. Pack slices (½ inch thick) in rigid containers, leaving ½-inch expansion space.

Onions

Onions of one sort or another are inexpensively available throughout the year, thank goodness, and so freezing them becomes more a matter of convenience than economy or out-of-season treasure, except for the sweet Bermudas and Spanish varieties.

Bermuda and Spanish onions may be frozen in raw slices for salads if great care is taken in packaging them and they are served while they are still noticeably frosty. The product is not nearly as good as the fresh, however, and freezing them is recommended here only because the flavor, especially of Bermudas, is like no other onion in the world, and the season is short.

To freeze them raw, cut them in ⅛-inch to ¼-inch slices and arrange the slices singly on a strip of cellophane, about two inches apart. Put another piece of cellophane over them and fold the covered strip accordion style. Wrap the mound of onion slices in heavy laminated freezer paper, or in heavy aluminum foil, and seal the package firmly with freezer tape. Put the package, or several packages, in a double polyethylene bag or in a rigid plastic or waxed tub container.

Chopped onions for cooked recipes are best frozen after blanching in water or steam just long enough to turn them limp and shiny (about two minutes). Chill them thoroughly in ice water, drain, and package in rigid containers or bag-lined boxes which are then wrapped in freezer paper and sealed.

Boiling onions, the little white ones that are so good creamed for Thanksgiving menus, may also be frozen after brief blanching (three to four minutes), chilling and draining, or they may be frozen cooked until tender. Be sure to label the package as to how long they have been heated, to guide you later. Pack in lined or polyethylene bags inserted in boxes and then wrapped in freezer paper.

Note: Onions require extra precautions in packaging not in self-defense, but to make them socially acceptable to the other foods in your freezer.

Parsley

This is a Complete Book about Freezing, and so I will tell you how to freeze parsley, which is of course a highly nutritive green. If you take my advice, however, you won't bother. When your garden bestows on you more parsley than you can use fresh, why not dry it slowly in a low oven and pack the flakes in envelopes or airtight containers for use as flavoring? If you insist on freezing it, here's how:

Drop fresh, curly, bright-green parsley tops in boiling water and let it stay there for no more than 20 seconds. Cool it immediately in ice water, and drain it thoroughly on absorbent paper towels. Package meal-size portions in cellophane, polyethylene or laminated bags. Seal the bags tightly. Put several bags in another large bag, or in a vapor-moisture-proof rigid container. Seal the container.

Parsley frozen in this way can be used in most cooked recipes calling for the green. It is not satisfactory in salads.

Parsnips

This is a hardy vegetable which takes well to freezing and can be used throughout the year as a side dish as well as in soups and stews, to which it contributes a good deal of delicate flavor.

Best varieties for freezing: American, Hollow Crown, Marrowfat.

Nutritive value: Fair source of calcium, phosphorus, iron and Vitamin B-1.

Approximate yield:

1 to 1½ pounds equal 1 pint.

1 bushel (45 pounds) equals from 30 to 40 pints.

Preparing for the freezer: Because of their high crude fiber content, parsnips should be harvested or purchased while they are young and tender. Wash, scrape and slice or dice parsnips into thicknesses no greater than ½ inch. It is not advisable to freeze parsnips whole, not even tiny ones. However, if your family likes french-fried parsnips as a vegetable, select the smallest and tenderest of the lot and cook them whole or halved according to the instructions for french-fried eggplant, page 248. After they are thoroughly cooled, package them in rigid containers or in laminated bags.

Blanching in water: 2 minutes.

Blanching in steam: 3 minutes.

Cool quickly and drain as completely as possible.

Packaging: Pack in polyethylene or laminated bags, or in rigid containers, leaving ½-inch expansion space.

Oyster Plant (Salsify)

Handle these according to the directions for Parsnips.

Peas

These best-sellers of the supermarkets can be equally successful in your cabinet if chosen and frozen as carefully by you as they are by the experts in the frozen food industry. As one of the eminent authorities-in-charge-of-research at the country's largest frozen food company told me, the reason why the home product frequently does not match the commercial one is that housewives attempt to freeze peas which are too mature. A single day's difference in harvesting may mean the difference between a firm, flavorful pea and one which is starchy.

If you grow your own, do be sure to gather them for freezing when they are not yet fully matured. If you buy them in quantity from a nearby farm or market, follow the suggestion for brine flotation at the end of this chapter to winnow out the not-quite-right ones.

Best varieties for freezing: Thomas Laxton (by far the best), Alderman, Little Marvel. Other good varieties are Freezonian, Laxtonian, Prosperity, Laxton's Progress.

Nutritive value: Good source of calcium and phosphorus, and high in the extremely volatile Vitamin C.

Approximate yield:

2 to 2½ pounds (in pod) equal 1 pint.

1 bushel (30 pounds) equals from 12 to 15 pints.

Preparing for the freezer: Shell as you ordinarily do; or, for easier shelling, plunge pods in boiling water for 1 minute then cool at once in ice water. For *still* easier shelling, do what the students do at Long Island Agricultural and Technical Institute: run them through an old-fashioned roller-type clothes wringer! Feed them through (after 1-minute blanching) stem ends first. Place a pan or colander between you and the wringer, for the pods will go through but the peas will drop on the near side.

Discard hard, wrinkled or damaged peas.

Blanching in water (preferred method):

1 minute if shelled raw.

½ minute if shelled after boiling pods.

Lift blanching basket in and out of water to distribute the heat evenly over the layers of peas. Plunge blanched peas immediately into very cold water, using plenty of ice to maintain a temperature of 50°F. or below. Drain peas until they are dry.

Packaging: Pack peas in pint or quart rigid containers, leaving ½-inch expansion space. They may also be packed in laminated or polyethylene bags.

Special note to upright freezer owners: A splendid way to insure really fast freezing is to line a freezer plate shelf with ordinary waxed paper and spread the blanched, cooled, drained peas over the shelf only one layer deep. When they are completely frozen, pour them into rigid containers or bags for storing.

Peas, Blackeye

Best varieties for freezing: Crowder, Blue Goose.

Blanch in water for 2 minutes, otherwise follow preceding directions for Peas.

Peppers, Sweet Green

If for nothing else, I have blessed my freezer a hundred times for its kind hospitality to peppers, which I have been accused of using in everything except ice cream.

The pepper situation may be different in your locality, but around here there is a brief period during early autumn when they can be bought cheaply. Later on in the year, I have paid almost as much for a pound as I had paid for a bushel earlier. One year, too, there was some sort of a pepper blight which made the handsome green bells as scarce as uranium, and almost as expensive.

For my small vegetable garden, I purchase young pepper plants as soon as the local stores display them and wait impatiently for the bells to mature.

Best varieties for freezing: California Wonder, World

Beater, Windsor. (Harvest while they are still dark green.)

Nutritive value: Good source of Vitamin B-1, B-2 and C when green.

Approximate yield: Three good-sized peppers equal one pint.

Preparing for the freezer: I freeze peppers in two ways, generally—halved, for stuffing and baking, and sliced, for frying in garlic and olive oil or for use in stews and casseroles.

I must confess, too, to breaking my own rule about conservation of essential food elements when it comes to green peppers, for I freeze a few bags of thinly sliced rings for crisp salad garnishes without blanching. My excuse for this is that at the same time I freeze enormous quantities of dutifully blanched peppers whose enzymes are properly inactivated.

Cut off stems and caps and cut in half lengthwise to remove seeds and fibrous parts. Wash thoroughly and slice those you are not going to freeze as halves.

Blanching in water (preferred method):

Halves, 3 minutes.

Slices, 2 minutes.

Cool thoroughly, and drain for dry pack.

Packaging: Halves are packed most easily in laminated or polyethylene bags; slices in rigid containers, leaving ½-inch expansion space. I prefer this dry pack, but peppers may also be covered with a brine solution, 1 teaspoonful of salt to 1 cup of cold water, and frozen in rigid containers allowing 1-inch expansion space.

Peppers, Sweet Red

These are handled in much the same way as green peppers and you may, if you like, mix the colors before packaging and freezing them. I usually freeze the red ones separately, often treating them first to make pimientos of them.

Best varieties for freezing: Perfection, Ruby King, World Beater (ripened to redness).

Nutritive value: Essentially the same as Green Peppers.

How to Make Pimientos

Roast red peppers on hot coals in an open fireplace or outdoor grill, or on the tines of a long fork held over a gas-range flame. Let them char completely. Plunge in cold water and rub off the charred skins. Remove caps and seeds. Pack tightly in rigid containers and cover with brine solution (1 teaspoonful of salt to 1 cup of cold water) leaving 1-inch expansion space.

Before serving, allow the pimientos to thaw in their containers. Pour them into a bowl or jar and cover them with their own brine solution to which you have added about 1 tablespoonful of olive oil, the expressed juice of 1 clove of peeled garlic and a few peppercorns. Or serve them plain.

When I am planning an all-Italian meal for a few of my slim friends who never seem to gain an ounce, I use these pimientos in a true Neopolitan antipasta: Alternate layers of pimiento, flat anchovies and paper-thin onion rings, sprinkled with olive oil and oregano (or basil, if you prefer this herb) and garnished with capers.

Potatoes, Sweet

These must be frozen cooked.

Scrub (with a brush, if necessary to get them clean) large- and medium-sized sweet potatoes that are fully mature. Boil, bake or pressure-cook them until almost tender. Allow them to stand at room temperature until they are cool. Peel them. Leave small potatoes whole, cut larger ones into halves, slices or wedges.

If you want to freeze mashed sweet potatoes, allow them to cook until they are completely tender, then scoop them out and mash them or put them through a ricer, adding about two tablespoonfuls of orange or lemon juice for each quart of mashed potatoes. (This prevents darkening.)

Dip whole, halved or sliced sweet potatoes in a mild solution of ascorbic-citric acid (see page 279) or in a quart of water to which you have added ½ cup of lemon juice. Let the potatoes remain in this solution for about 5 seconds, making sure that the liquid has fully covered them. Drain the potatoes on paper towels and package them just as they are in laminated bags or bag-lined boxes, or roll them first in brown sugar in readiness for later candying.

Pack mashed sweet potatoes in rigid containers, leaving ½-inch expansion space.

Potatoes, White

Heavy plastic bags of uniformly sized peeled raw white potatoes, ready to cook or use in recipes, are available in store freezer cabinets. I have used them. I do not, however, freeze my own uniformly sized peeled raw white potatoes because I do not have facilities for the flash-freezing techniques perfected by the frozen food industry.

Nor do I freeze French-fried potatoes, although I could. The industry does, with marked success, and I sometimes buy a big bagful to have on hand in the freezer for quick serving—but never to a certain young man of my acquaintance, now a teen-ager. When he was six or seven years old, I served him French-fried potatoes of my own customary preparation—hand-peeled, hand-cut into fingers, soaked in ice water in the refrigerator for an hour or more, drained, dried between towels, then dropped into hot deep fat until golden brown. After he tasted a few, he asked what they were. It took some heart-crossing and hope-to-die to convince him he was eating French-fried potatoes. I learned he thought they came exclusively in freezer bags or boxes, and *these* tasted better than ice cream, even.

On occasion I do freeze white potatoes stuffed-baked for reasons of time-saving, fuel economy and convenience. It is just as easy to bake several extra nice large Idahoes when I'm baking the dinner portion, and little trouble to remove pulp from shells, mash with milk or cream and butter and a little salt, and return the pulp to the shells. When cool, these are wrapped in foil or film and frozen for heating at a later date.

Pumpkin

If you use pumpkin as a vegetable, prepare full-colored, fine-textured pumpkin as you would for the table—removing the seeds and either boiling, steaming or baking—then remove the pulp and mash it while it is hot. It may also be put through a potato ricer. Cool the mashed pumpkin by putting a bowlful in cold water, stirring the pulp occasionally. When it is cold, pack the pumpkin in rigid containers, leaving ½-inch expansion space.

You can add to your popularity by freezing enough of your favorite pumpkin pie mix to keep your family happy all year. Prepare this as you always do, being sure to use homogenized pasteurized milk or heavy cream. When it is cool, package enough for one pie in a rigid container, or put it in unbaked pie shells.

Rhubarb

I know this is a vegetable, but if you don't mind I am going to discuss it in the chapter on Fruits.

Rutabagas

See Turnips.

Spinach

The spinach you freeze should be of the deepest imaginable green, and should be picked while the leaves are young and tender, before the appearance of seedstalks.

Best varieties for freezing: Savoy Leaved, Giant Thick Leaved, Long Standing Bloomsdale, Nobel, King of Denmark, Old Dominion, Viking, New Zealand.

Nutritive value: Good source of iron, manganese, Vitamins A and B-2.

Approximate yield:

1 pound or a little more equals 1 pint.

1 bushel (18 pounds) equals from 12 to 18 pints.

Preparing for the freezer: Remove large stems and any bruised or discolored leaves. Wash with a sink-spray or under running water for a long, long time to get rid of every last vestige of sand and dirt. The Long Island Agricultural and Technical Institute uses a washing machine for de-gritting spinach—in warm water, to decrease the time it takes the blanching water to return to 212°F. Work with small amounts, no more than two pounds at a time, to prevent nutritive loss. Pull large leaves apart, or cut them into pieces with scissors.

Blanching in water (only method recommended): Boil for 2 minutes in at least 3 gallons of water for each pound of spinach. If you are freezing New Zealand spinach, blanch in boiling water for only 1 minute.

Cool quickly under running, cold water. Do not let it mat together. Drain spinach as much as you possibly can, but don't expect ever to drain it dry. Somehow, you never can. You think it's as dry as blotting paper, then all of a sudden it breaks out in little droplets of dew. You can remove some excess moisture by pressing the leaves with a wooden spoon.

Packaging: Pack loosely in laminated or polyethylene bags or in rigid containers.

Note: If what you've got is spinach of whose age you are not certain, but you suspect maturity, blanch it for two minutes and then run the leaves through your food chopper. Add onion seasoning, if you wish. Package in rigid containers as it comes from the chopper, or add a light cream sauce (using homogenized, pasteurized milk) and freeze in rigid containers, being sure to leave at least ½-inch expansion space.

Squash, Summer (also *Zucchini*)

Best varieties for freezing: Prolific, Straightneck, Table Green, Cocozelle.

Nutritive value: Not an impressive source of minerals or vitamins, but a good low-calory vegetable.

Approximate yield:

1 pound equals 1 pint.

1 bushel (40 pounds) equals from 32 to 40 pints.

Preparing for the freezer: Select for freezing summer squash or zucchini whose rind is still a little tender, and with small seeds. Wash under running water, and cut, without peeling, in ½-inch slices.

Blanching in water: 3 minutes.

Blanching in steam: 4 minutes.

Cool immediately in very cold water and drain on absorbent towels.

Packaging: Pack without pressing down in rigid containers, leaving ½-inch expansion space.

The best way to use frozen zucchini, I have found, is to dip it in crumbs, egg and crumbs again, and fry in deep fat. An alternate and good method for freezing zucchini is to french-fry it in the manner described for Eggplant and freeze it pre-cooked.

Squash, Winter

Best varieties for freezing: Delicious, Buttercup, Golden Hubbard, Green Hubbard, Acorn.

Nutritive value: Excellent source of Vitamin A.

Approximate yield:

3 pounds equal 1 quart.

Acorn squash—1 squash equals 2 servings.

Preparing for the freezer: Prepare the larger or long-necked varieties in much the same way as you do pumpkin, by washing and removing the seeds and cooking the pieces until soft. Remove pulp from the rind, and rice or mash. Cool in a bowl placed in cold water, stirring occasionally. Pack in rigid containers, leaving ½-inch expansion space.

Acorn squash may be prepared the above away, or—as I prefer it—cut into halves and baked with butter in a medium oven until tender. Allow the halves to cool at room temperature. Wrap and seal each half in cellophane or aluminum foil, and package several together in large polyethylene bags. A grand vegetable served alone, or use them occasionally as shells to contain mixed seafoods.

Succotash

Follow individual directions for preparing Cream-Style Corn and Baby Lima Beans. Mix them in equal proportions and pack in rigid containers, leaving ½-inch expansion space.

Swiss Chard

Best varieties for freezing: Lucullus, Fordhook, Green Plume.

Prepare exactly as you do Spinach.

Tomatoes

If, like me, you use a lot of cooked tomatoes and dislike canning, you will want to freeze some whole. You may want to freeze tomato juice, too. And if you own an upright freezer, you can actually freeze them raw for salads.

Almost any garden tomato can be frozen successfully, although those low in acid are best for juice; small ones are best for whole cooked tomatoes, and small, almost seedless, ones picked early in the season are best for salads.

Nutritive value: Fair source of Vitamins A and C.

Preparing for the freezer:

Whole cooked tomatoes. Wash, remove stems and any spots. Put tomatoes in a flat pan of rapidly boiling water until their skins crack. Remove with slotted spoon and let cool until they can be handled. Remove skins and cores and simmer in their own juice until they are soft but not falling apart (about 5 to 7 minutes). Cool at room temperature, then chill in refrigerator. Package in rigid containers, covered with their own juice, leaving 1-inch expansion space. Use as a cooked table vegetable, or in sauces, soups and stews.

Tomato juice. Wash, trim, core and quarter ripe tomatoes. Simmer them covered for about 10 minutes. Press through a fine-meshed sieve or cheesecloth. Allow to cool and pour into rigid containers or shoulderless glass jars, leaving 1-inch expansion space. To serve, allow juice to thaw slowly in the refrigerator. When it is liquid, add salt, pepper, lemon juice and, if you like tomato juice this way, a dash of tabasco sauce. Stir well.

Whole raw tomatoes for salads. NOTE: This method is usually successful only in upright freezers with freezer plate shelves or in chests which have fast-freezing compartments.

Handling carefully to prevent bruises, place unwashed whole tomatoes directly in contact with a freezing plate. (You may line the plate first with cellophane or freezer paper.)

When the tomatoes are frozen you will notice that the skins are cracked in several places and easy to remove. Peel the tomatoes rapidly and pack them immediately in laminated or polyethylene bags and return them to the freezer for storage. Put them in a corner or in a place where they will not be crushed by heavy packages.

Before you plan to use them, take out the number desired and let them thaw partially in your refrigerator. They should be served in salads while they are still frosty, before all the ice crystals have disappeared.

Turnips and Rutabagas

Best varieties for freezing:

Turnips: Purple-top White Globe, Strapleaf.

Rutabagas: Long Island.

Nutritive value: Good source of iron and manganese; fair source of Vitamins B-1 and B-2.

Approximate yield:

1 to 1½ pounds equal 1 pint.

1 bushel (50 pounds) equals from 35 to 45 pints.

Preparing for the freezer: Select young, tender, small to medium-size turnips or rutabagas. Cut off tops, wash, peel, and cut into ½-inch cubes.

Blanching in water: 2 minutes.

Blanching in steam: 3 minutes.

Cool immediately in cold water and drain thoroughly. Pack in rigid containers, leaving ½-inch head space.

Rutabagas and yellow turnips may also be frozen mashed. Cook until tender in boiling water, drain, rice or mash. Cool by placing bowlfuls in cold water, stirring occasionally. Pack in rigid containers, leaving ½-inch expansion space.

Turnip Greens

Nutritionally speaking, these are the best parts of turnips and are especially high in minerals as well as in Vitamins A and C. Prepare turnip greens as you do Spinach, blanch in water for 2 minutes, pack loosely in bags or containers.

Weeds

Yes, weeds. Those impudent intruders over which you break your back in the garden, muttering unlady-like words as you throw them away or burn them, are good eating by savory standards as well as nutritional. The next time you do your weeding chores, it might be interesting for you to discover which of the edible greens you have been rudely chucking out. A good book on gardening will show you pictures of the following list, compiled in the interests of nutrition by the eminent scientist, Dr. George Washington Carver:

Dandelion greens, ox-eye daisy leaves, wild lettuce, hawk-weed, curled or western dock, sorrel, shepherd's purse, stock, careless weeds, pokeweed.

If you feel in an experimental mood, prepare any of these for freezing just as you handle Spinach. When serving, they may be eaten alone or mixed with other greens such as swiss chard, spinach or turnip greens.

Try milkweed pods, too, handling them according to the directions for Okra.

Zucchini

See Summer Squash.

Brine-Flotation Test for Lima Beans and Peas

If you approach the freezing of these two vegetables with determination to obtain the nearest-to-perfect pack possible, you may want to subject "run-of-pod" limas and peas to the flotation test which has been reported successful by the Home Economics Division of the University of Maryland.

Green Lima Beans

Dissolve two full measuring cups of free-running salt in 2½ gallons of water whose temperature is 70°F. Add to this solution (about two pounds at a time) shelled, washed lima beans. Stir and allow the beans to settle, skimming off the floaters and those suspended in the solution with a slotted spoon or a small kitchen strainer. These, the floaters and swimmers, are the choice limas for freezing. Wash them again, discarding off-color or split beans, before proceeding with blanching according to directions.

Peas

Do the same thing with peas which have been shelled, washed, *blanched and cooled,* but using two full measuring cups of salt to only 1 gallon of water.

In this test, too, the floaters and swimmers are choice for freezing.

In both cases, the "sinkers" are not to be discarded by any means. Cook these for immediate table use, or freeze them separately, marking the packages as "second quality" so that you can compare the difference in taste and texture with the choice group at a later date.

FREEZING VEGETABLES FROM A TO Z

DESCRIPTION OF VEGETABLES FOR FREEZING, AND SEASONS OF YEAR WHEN MOST PLENTIFUL

Vegetable	*Description*	*In Season* *
Artichokes		
Globe, French or Italian	Round or globular; resembles tight green bud; choose small ones.	Nov.-May
Jerusalem	Looks like a knotty potato. Not for freezing.	
Asparagus	White or green.	April-June
Beans		
Green or wax	Round or flat, no longer than 5 inches long.	All year. First harvest of each sowing best for freezing
Lima	Flat, kidney-shaped green or white seeds in green pods.	Aug.-Sept.
Beets	Choose small, tender varieties for freezing. Large beets may be dry-stored in root cellars.	March; May-Dec.
Broccoli	Sprouting (flowret, rather than compact heads) preferred for freezing.	Nov.-May
Brussels sprouts	Choose small to medium sprouts.	Oct.-March
Cabbage	Young, tender heads preferred.	All year
Carrots	Choose small, tender, almost coreless varieties for freezing.	All year
Cauliflower	Firm, compact heads.	Sept.-June
Celery	Freeze pre-cooked only.	Oct.-April
Collards	Resemble loose-leaved cabbage.	Local harvest
Corn	Small to medium yellow varieties best for freezing on cob; use white types for cream style.	July-Sept.
Cucumbers	Freeze in vinegar.	April-Sept.
Eggplant	Semi-mature varieties with tender seeds.	Available all year. Best months Mar.-June and Aug.-Oct.
Kale	Tightly curled, tender leaves.	All year
Kohlrabi	Turnip-like surface tuber; select young, small bulbs with tender leaves.	Best months: Aug.-Sept.
Mushrooms	Cultivated varieties, preferably white or cream-tinted caps.	In best supply: Oct.-Jan.

DESCRIPTION OF VEGETABLES FOR FREEZING, AND SEASONS OF YEAR WHEN MOST PLENTIFUL (*Continued*)

Vegetable	*Description*	*In Season* *
Okra	Long, green, fuzzy pod.	July-Oct.
Onions		
Bermuda		May-July
Spanish		July-Oct.
Parsley	Curly varieties.	May-Nov.
Parsnips	Young, tender roots chosen before they become too woody.	Oct.-April
Peas	Young, tender seeds harvested before turning starchy.	All year; best when fresh-picked in local markets
Peppers	Green or red.	Varies with region. Best months usually Aug.-Oct.
Potatoes		
Sweet	Freeze after cooking; medium to large roots.	Sept.-March
White	Freeze french-fried or baked-stuffed only.	All year
Pumpkin	Choose for fine texture; avoid stringy varieties.	Sept.-Dec.
Rutabagas	Purple-top varieties.	July-April
Soybeans	Firm, plump, bright green pods.	Regional harvest
Spinach	Deep green, tender leaves.	Oct.-June
New Zealand		June-Nov.
Squash		
Summer (Zucchini)	Tender rinds; small to medium, with small seeds.	May-Oct.
Winter	Select mature varieties for fine texture.	After frost
Swiss chard	Dark green, curled leaves.	Local harvest
Tomatoes	All early and late varieties with good color; choose small, almost seedless fruit for freezing whole.	April-Nov.
Turnips	Small to medium roots.	Sept.-March

* Seasonal information compiled from statistics furnished by *Chain Store Age*, a grocery trade publication. This is a national average. For the peak season in your own region, consult your county agent.

Always remember to place packages against the side walls of a chest freezer, in a single layer on freezer plate shelves or in fast-freezing compartments an inch or so apart, in order to allow air circulation between the packages.

All of the blanched raw vegetables discussed in this chapter will retain their taste, quality and texture for periods of time up to 10 months at 0°F., to 6 months at 5°F. Use unblanched pepper rings or tomatoes within 3 months, and do not store cooked vegetables such as french-fried egg-plant, etc., longer than 4 months at 0°F. or 2 months at 5°F.

17. Fruits for Freezing

One of the most rewarding services performed by your home freezer is also one of the easiest for you to enjoy, for it will preserve delicious tree- or vine-ripened fruits, to be savored in all their garden goodness long after the garden itself is barren and blanketed with snow.

Fruits require little more preparation for the freezer than they do for immediate table use. Unlike vegetables, they do not need to be subjected to high heat (blanching) in order to insure palatable and nutritionally protected results. Freezing itself, and the sugar in dry or syrup form which you will add to most fruits, retard enzymic action to an appreciable extent.

Why Some Fruits Darken—and How to Prevent It

You are familiar with how quickly a peach, pear, banana, apple or other light-fleshed fruit will turn unappetizingly dark soon after its skin has been broken. The skin or peel of a fruit is nature's inimitably successful protective packaging material, shielding the flesh against insects, bacteria, molds and the oxidizing effect of air itself.

Within the cells of fruit there are tiny little crystalline compounds, called *catechols* and *tannins* by the chemists who discovered their presence. In their natural and undisturbed state, safely imprisoned within the rind-protected fruit, these substances are colorless. The instant they are exposed to air, however, they celebrate (or protest) by becoming highly colored, and the result is darkened fruit which soon loses its fresh aroma and flavor.

While these temperamental catechols and tannins are present in varying degrees in all fruits, berries and melons, you have noticed that some species seem to resist the darkening influence better than others. For example, if you leave half an apple and half an orange on a plate, the orange will retain its original color long after the apple has turned almost black. This interesting phenomenon is caused by the different content of Vitamin C in each of the fruits.

Vitamin C is *ascorbic acid,* a substance which forms a natural barrier to air oxidation of the vulnerable crystallines by permitting itself to be oxidized instead. Air seems to like Vitamin C for its oxidizing purposes better than it likes the catechols and tannins. The more Vitamin C a fruit possesses, therefore, the longer it takes the air to get around to oxidizing, or darkening, the fruit.

There is still another deterrent to darkening of light-fleshed fruit, and this is *citric acid.* Oranges, lemons, limes and grapefruit, generically called *citrus fruits,* are generously supplied with citric acid, as well as with ascorbic acid (Vitamin C). When air, poking around and looking for catechols and tannins to darken, runs across Vitamin C *and* citric acid, it spends its time oxidizing first one and then the other, and meanwhile the browning susceptibles within the fruit remain unmolested.

This fascinating chemical life process of fruits is an important factor in freezing procedure. It explains why you must, first of all, work very quickly when you are freezing fruits, why many fruits are best when a form of ascorbic-citric acid is added to the syrup, and why fruits should be served and eaten immediately upon thawing or while a few ice crystals still remain in them. Fruit purées, especially, should be protected against darkening, for large quantities of air are forced into the pulp as it is crushed, sieved or put through a food mill.

You can have a lot of fun freezing fruits, and the results of your comparatively simple labors will be extremely profitable. You can freeze some whole fruits and berries, pie fillings, mixed fruits for salads, puréed fruits for sauces, crushed fruits for cold drinks or for making into canned jams and jellies in the wintertime, when canning is not such a chore. You can freeze your own fruit (and vegetable) juices, too, a handsome variety of them to relieve breakfast boredom.

You can do surprising and creative things with fruits and a freezer. Last Christmas season, for example, my drop-in guests were delighted when they were given frosty sherbet glasses piled high with Christmasy balls of mixed red watermelon and green honeydew.

There are only a few "musts" to remember when you decide to entrust to your home freezer a variety of summer's fruits for later enjoyment. If you are one of the lucky ones and have your own fruit trees, melon vines and berry bushes, be patient and wait until the fruit is completely and deliciously ripened. The exception to this is elderberries, which seem to be higher in Vitamin C when slightly unripe. Pick fruit or berries in the cool of the morning and freeze them the same day.

If your fruit is bought from a farm or market, buy only as much as you can handle in a single day.

Before you start working, have plenty of ice on hand. Make blocks and cubes of ice in your freezer the day before. Fruits and berries must be washed, and ice water firms them quickly, preventing sogginess and loss of juices.

Use only those utensils which are made of enamel, earthenware, aluminum or stainless steel when preparing fruit. Iron, copper or thinly tinned vessels may produce off-flavors, and galvanized ware must be avoided like the plague because the zinc used in manufacture dissolves in fruit acid juices.

Drain fruits, when draining is called for, in a colander or on absorbent paper towels, taking care not to jostle them too much.

Use a huller, knife or the tips of your fingers to remove the stems and caps of berries. Never squeeze them off. Work with only a quart of berries at a time, or two or three quarts of fruit.

The packaging materials you use for fruit must be firm, vapor-proof and absolutely watertight, for some fruit juices are so acidic that they may eat their way through insufficiently treated paper or cardboard, making a mess in your nice white cabinet and leaking over other food packages. As a rule, I prefer to use glass jars or plastic containers for cherries and juicy berries, but have used the laminated bag-and-box combinations as well as waxed cardboard tubs for other fruits and dry-packed blueberries.

As with all other frozen foods, pack only as much of any

fruit in a container as will be consumed at one meal, or for your anticipated amount of pies to be baked at one time. A quart of berries, for example, will make four generous servings as a dessert; a pint will do as a sauce for four; a quart of fruit makes a full 9-inch pie.

Most important of all, of course, is the fact that fruit or berries to be frozen must be of excellent quality if the end product is expected to be good. Your freezer preserves the original goodness of food, but it does not perform feats of magic.

Four Ways to Pack Fruits for Freezing

Depending on the fruit itself, the use for which it is intended and the degree of sweetness desired in the ultimately defrosted product, you will ordinarily make a choice among four acceptable methods of packing. These are the Unsweetened Dry Pack, the Unsweetened Wet Pack, the Dry Sugar Pack, and the Syrup Pack.

1. UNSWEETENED DRY PACK

This method is used primarily for fruits or berries which are intended for pies, jellies, jams or preserves or for people on special diets.

Procedure: Wash fruit or berries, drain them in a colander, and pack immediately in rigid containers. Seal and freeze at once. Be sure to mark containers as to the type of fruit or berry and add "Dry, Unsweetened" to the information on the label.

2. UNSWEETENED WET PACK

Wash and prepare as for the table any fruit or berry you wish to use without sweetening, either for dietetic reasons or because you may decide the fruit is already so sweet with its own natural sugar it would be an insult to add more.

Procedure: Pack prepared fruit in liquid-proof containers, either crushing them gently in their own juice or covering them with water which contains an ascorbic-citric acid solution. (Directions for using the ascorbic-citric acid treatment will be given a little later.) Allow at least ½-inch expansion space before putting the lid on the container. If fruit is tart but you wish to avoid using sugar, try covering it with any favorite flavor of a sugarless carbonated beverage or a solution made with water sweetened with a sugar substitute. Label the package completely, seal and freeze at once.

3. DRY SUGAR PACK

Procedure 1: Without treatment by ascorbic-citric acid, not necessary for most varieties of berries, which apparently contain sufficient amounts of the natural anti-oxidants to prevent their discoloration and loss of flavor.

Prepare fruit or berries carefully, as you would for the table. Put the prepared fruit or berries in a large bowl or shallow pan, working with one quart or less at a time. Sprinkle over them sugar in the amounts recommended in the directions for each fruit. Use a slotted spoon (preferably wooden) or a steel or aluminum pancake turner to lift the fruit through the sugar until it is evenly coated, and continue mixing gently until the juices are drawn out and the sugar is dissolved. Pack in rigid containers or laminated bags in boxes, label accurately and freeze at once.

Procedure 2: With ascorbic-citric acid treatment, recommended for fruits which darken quickly, such as peaches.

Prepare the fruit as for the table. Prepare treated sugar by mixing 3 level teaspoonfuls of ascorbic-citric acid powder with every 2 cups of sugar used, mixing the ingredients thoroughly. Sprinkle the treated sugar on the bottom of the container, then fill container one-fourth full and sprinkle more treated sugar over the fruit, using a little less than one-fourth of the total amount called for in the instructions. Next, fill the container to the halfway mark with fruit and repeat the sugar treatment. Continue with this alternating procedure until the container is filled to within ½ inch of the top; sprinkle the top fruit with treated sugar, too. Seal, label and freeze at once.

4. SYRUP PACK

Syrup for freezing fruits may be made of water and sugar, water and white corn syrup, water and honey or water and a combination of all three, according to your family's taste.

Once, in an inspired frenzy of nutritional pioneering, I experimented with a syrup made of water and blackstrap molasses, which I used to cover peeled pear halves. It was a perfectly good preservative, I discovered, but I was the only one who would eat the odd-looking dark gray pears, feeling virtuous and a little silly as I did so.

I mention this because there are many people who consider refined white sugar to be a dietary menace when consumed in quantity. I have asked several of the commercial packers if, for these people, brown sugar could be substituted in the syrup pack. It can be, if the color doesn't bother you. Honey, too, can be used by those who disdain white sugar, but syrup made with honey in sufficient quantities to be a preservative is likely to make the fruit taste more like honey than fruit.

The syrups used for fruits and berries run the gamut from very thin to very heavy, and which one you use will depend on how sour the fruit is, and how sweet you want it to be.

Always prepare your syrup at least a day before you are going to use it, and keep it in the refrigerator until you are ready for it. Its temperature should be well below 70°F. for best results.

For those fruits which you are going to protect further against discoloration, add the ascorbic-citric acid powder to *cold syrup* (never warm or hot) just before using it.

Ascorbic-Citric Acid

While ascorbic acid may be used alone, it has been found that a mixture of ascorbic and citric acids is somewhat better—and cheaper, too, for ascorbic acid alone is a little high in price. Citric acid acts as a stabilizer for the ascorbic acid (Vitamin C) and extends the length of time the fruit is protected against oxidation.

Any good pure-food or pure-drug company's ascorbic-citric acid preparation may be used, and I believe there are several on the market. For my purposes, I prefer the ascorbic-citric mixture ("A.C.M.") put out by Chas. Pfizer & Co., Inc., for I have a high regard for this century-old firm and its intensive research in frozen food problems as well as in the fields of pharmaceuticals and medicinals.

The amounts of acid given in directions for freezing fruits are based on the A.C.M. product. If you use other products, follow the directions given by the manufacturers.

The following tables will tell you the proportions of sugar, corn syrup and water to use in making the various percentages of syrup mentioned in the directions for freezing each variety of fruit. The strength of the syrup recommended varies with the type and tartness of the fruit, and the strengths given are proportioned to the average palate. You may like your fruits less or more sweet than average. If so, just go up and down the scale of percentages for the strength of your preference.

How to Make Syrup for Freezing Fruits

SUGAR SYRUP: *Use boiling water*

Per Cent of Syrup	*Cups of Sugar*	*Cups of Water*	*Yield: Cups of Syrup*
20	1	4	5
30	2	4	5⅓
40	3⅓	4	5½
50	4¾	4	6½
60	7	4	7¾
70	9	4	8⅔

Dissolve the sugar completely in boiling water, wait until the syrup cools and then store it in the refrigerator until you are ready to cover the fruit for freezing. Whenever possible, a 40 per cent syrup—or less—is preferred, as heavier syrups tend to make fruits flabby.

Ascorbic-citric acid powder is added to cold syrup. Follow manufacturer's directions for amount to use for each fruit.

CORN SYRUP WITH SUGAR: *Use cold water*

Thin syrup:

1 cup sugar, 2 cups corn syrup, 6 cups water, (yield) 8 cups syrup

Medium syrup:

2 cups sugar, 2 cups corn syrup, 5 cups water, (yield) 8 cups syrup

Heavy syrup:

3 cups sugar, 2 cups corn syrup, 4 cups water, (yield) 8 cups syrup

Dissolve the sugar first in the measured amount of *cold*

water called for, then add the corn syrup and mix thoroughly. Store in refrigerator until needed.

Honey may be substituted for the corn syrup, if desired. Remember, however, that both corn syrup and honey may give a foreign flavor to the fruit.

Ascorbic-citric acid powder is added to cold syrup. Follow manufacturer's directions.

How Much Syrup Should You Make?

This may seem like an academic question, but once or twice I found myself at the quart-end of a bushel of fruit without any syrup to cover it with. It can happen. And, because fruit must be covered with *cold syrup,* it meant that I had to hold the fruit over while I went through the motions of making more syrup and chilling it. No mathematician I, nevertheless I had to wet the end of a pencil on my tongue and do some figuring. Using my mental gyrations as a basis, you can work out the amount of syrup needed for any given pack of fruit, as follows:

Problem: To freeze a bushel of peaches in medium (40%) syrup.

Calculating from the "approximate yield" information given for each variety of fruit, I knew that one bushel of peaches would yield anywhere from 32 to 48 pints, depending on how many of the peaches it would be necessary to discard because of bruises or immaturity. Striking a brilliant average of the two extremes, I deduced that I would have 40 pints of peaches from my bushel.

Now, between one-half to two-thirds of a cup of syrup is needed for each pint package of fruit, I knew by experience. Therefore, for 40 pints of peaches I would need from 20 to 27 cups of syrup. To be on the safe side, I decided to make 27 cups, because the excess amount could always be held in the refrigerator for a few days.

I knew that 3 cups of sugar and 4 cups of water would give me 5½ cups of 40 per cent syrup (see table on page 280) and 27 cups of syrup is a little less than five times 5½ cups, so I multiplied the measures for both sugar and water by five. Fifteen cups (7½ pounds) of sugar and 20 cups (10 pints) of water actually did give me enough 40 per cent syrup to accommodate 40 pints of peaches, and I found myself wishing again that I had paid more attention in my arithmetic classes. I'm sure there must be an easier way to arrive at a correct answer!

Which Method of Packing Fruit Is Best?

As I mentioned earlier, the degree to which you recognize your own and your family's sweet-tooth is one guide. Another is the tartness of the fruit or berry itself. A third is the purpose for which the fruit is to be used.

Generally speaking, very juicy fruits and berries can be successfully frozen in a dry pack, for the juices will emerge to form their own syrup with the dry sugar. I would rather not lay down any rigid rules for fruit processing, because your own ingenuity and experimentation can lead you to some very interesting results with almost any fruit.

One last word: Be sure to leave enough expansion space at the top of rigid containers to prevent breaking or splitting of packages when the liquid content expands during the freezing process. For liquid packs, allow ½ inch for pints, 1 inch for quarts. If the containers you use have narrow top openings, allow 1 inch for pints, 1½ for quarts. Dry packs require only ½-inch head space for any size or type of container.

Always prevent fruit from rising out of the syrup by crumpling a piece of cellophane or parchment paper over the top before closing the container.

Freezing Fruits

Apples

Apples are to be commended for their freezing versatility. They may be sliced for pies, fruit cocktails, cobbler or brown bettyish desserts; they may be pressed for a highly nutritious and tangy breakfast juice; they may be baked whole and frozen; and, when you have more apples than you know what to do with, they may be made into an exceedingly flavorful applesauce which is far superior to the canned product.

Best varieties for freezing: Largely dependent on whether apples are to be used for baking, sauce or juice, and frequently dependent on regional produce. As a general rule, firm winter varieties are most suitable for the freezer.

Approximate yield:

1 to 1½ pounds equal 1 pint.
1 bushel (48 pounds) equals from 32 to 40 pints.
1 box (44 pounds) equals from 29 to 35 pints.

Apples for pie:

Select fully ripened, crisp, firm-fleshed apples which you ordinarily use for pie-making. Have a large pan of very cold water at hand and, after washing, peeling and coring apples, let them bob around in the water until you slice them. Slice medium-sized apples into twelfths, large ones into sixteenths. Pack either dry and unsweetened; sweet dry (1 cup of sugar to 3 cups of apples), or in a 40 per cent syrup treated with ascorbic-citric acid.

If you use the syrup pack, slice the apples directly into containers which are one-fourth full of syrup, adding more syrup as you go along to keep the apples covered.

Another way of treating pie apple slices is to steam blanch them for 2-3 minutes (see directions for steam blanching on page 224) and then cool them quickly in ice water before packing.

Preferred method: Unsweetened or sweetened dry pack with ascorbic-citric acid treatment.

Apples for fruit cocktail or desserts:

Prepare as for pie, but the preferred pack is in a 40 per cent syrup treated with ascorbic-citric acid.

Apples, baked:

Use fairly large, firm apples. Cortlands are excellent, if you can get them. Other good varieties for baking are Northern Spies, Baldwins, Rome Beauties, Jonathans, York Imperials, Winesaps.

Wash apples thoroughly, and stem them. Remove the core to within about ½ inch of the bottom (blossom end) so that the apple will retain whatever sweetening and flavoring mixture you put in it during the baking process.

Fill the hollow with your own favorite mixture or with one of the following: brown sugar and cinnamon (½ cup of sugar and ½ teaspoonful of ground cinnamon for 6 apples); maple syrup and cinnamon; sugar or maple syrup with nutmeg and a little lemon or orange juice. Bake the apples in an oven pre-heated to 400°F. until they are tender. Cool them at room temperature.

Packing baked apples is a little tricky unless you are willing to devote a small waxed tub container to each one. They may be put into foil- or cellophane-lined candy boxes, separating each apple from its neighbors with more cellophane, and the closed and sealed box then wrapped and sealed again in laminated freezer paper.

A lazy but effective way is to fast-freeze a cookie tray of baked apples on an upright freezer shelf or in the fast-freezing compartment of a chest cabinet. When frozen they're a cinch to handle and may be individually wrapped in foil or cellophane and put into a large polyethylene bag, six or a dozen together.

Apple juice:

If there is an apple press in your community or up in the country where you spend week-ends, bring home gallons of the stuff, keeping it as cool as possible until you get home to pour it into plastic or glass containers for freezing. Add about 1 teaspoonful of ascorbic-citric acid powder for every gallon of juice.

Don't try to freeze store-bought apple cider, however, because it usually has a preservative in it that does not take kindly to zero temperatures.

Be sure to leave at least 1-inch expansion space.

Applesauce:

Early apples may be used for sauce, as well as the later varieties, just so long as they are full-flavored. Wash the apples, and slice them. You may peel them if you like. I prefer to simmer mine unpeeled for added tang.

Add one cup of water to each three quarts of apples. If you like a looser sauce, make the proportions 2 to 3. Cook slowly over low heat until tender, then cool the sauce. Force it through a strainer or put it through a food mill to remove seeds, peels and fibers. Sweeten to taste.

To prevent darkening, add 1½ level teaspoonfuls of ascorbic-citric acid powder for each cup of sugar used.

Pack in rigid containers, leaving ½-inch expansion space, seal and freeze.

Apricots

This thin-skinned, delicately flavored fruit is usually packed in golden halves without peeling or in peeled slices, and makes a delicious dessert, pie or cocktail fruit. Peeled apricots may also be puréed for use as a fine-tasting cake sauce or in home-made ice cream.

Best varieties for freezing: California Blenheim, Tilton, Moorpak, Hemskirt, Royal.

Approximate yield:

Slightly more than ½ pound equals 1 pint.
1 crate (22 pounds) equals from 27 to 34 pints.
1 bushel (48 pounds) equals from 60 to 80 pints.

Apricot halves for uncooked desserts:

Select firm, fully ripened apricots that have smooth yellow-orange skins. Wash under cold running water, cut into halves and remove pits. If the apricots are left unpeeled, dropping the halves into boiling water for ½ minute will prevent the skins from toughening in the freezer. Chill in ice water and drain.

Syrup pack preferred: Drop cooled halves directly into rigid containers which are ¼ filled with ascorbic-citric acid treated 40 per cent syrup. Fill containers to within ½ inch of top and cover with the treated syrup. Seal and freeze.

Apricots, sliced, for uncooked desserts:

Drop whole apricots into boiling water for 1 minute. Lift out and plunge immediately into ice water until thoroughly cool. Skins are then easy to remove. Slice directly into containers as above, adding more treated syrup as you proceed to keep the apricots covered. Fill to within ½ inch of top, crumple cellophane or parchment paper over the apricots, seal and freeze.

Apricots for pies and cooked desserts:

These may be frozen in exactly the same way as for uncooked desserts, or the dry sugar pack may be used to save freezer space. Use 1 cup of acid-treated sugar for each 4 to 6 cups of apricots. Lift the apricots through the sugar with a slotted wooden spoon until all the fruit is covered and the sugar is dissolved.

Pack in rigid containers, pressing the fruit down until it is covered with its own juice. Place crumpled cellophane or parchment paper over the top of the fruit, seal and freeze.

Apricots, puréed:

Drop whole apricots into boiling water for 1 minute, and cool immediately in ice water. Peel and pit the apricots and cut them into halves or quarters. Force through a sieve or food mill and mix the pulp with acid-treated sugar, ½ cup or less of sugar for every quart of pulp. Pack in rigid containers, leaving ½-inch expansion space. Seal and freeze.

Avocados (Alligator Pears)

It is a pity that this fruit cannot be successfully frozen as it comes from the tree in its beautiful dark-shining green or purplish shell, for almost anything you do to it in order to preserve the flesh is bound to affect the subtlety of its taste.

In my adventurous experiments, I have frozen avocado halves without doing anything to them beyond removing the stones and sprinkling ascorbic-citric acid *solution* over the cut sides, then wrapping each half in polyethylene sheeting with an over-wrap of aluminum foil. I have also treated sliced avocado with acid and frozen it for salads. You may do these things, too, if you wish; I am not saying you should.

Although avocados are on the market all year round (when the California season is on the wane the Florida crop reaches its peak) there are certain times of the year when the markets apparently overpurchase or there is a superabundance of the fruit, whereupon the prices drop considerably. This is the time to buy up dozens and mash them up for mixing later with all sorts of flavorings. Avocado spread flavored with onion, garlic, cheese, anchovy paste or your favorite herbs is superb on crackers or toast rounds for cocktail canapes.

Avocados should be fully ripe, ideal for eating. Select those whose shells are unbroken and free of scars. Cut in half, remove stones, scoop out meat close to the shell and mash thoroughly. Add 2 tablespoonfuls of lemon juice (or a solution of ½ teaspoonful ascorbic-citric acid powder to 1 ounce of water) for each avocado. If you prefer it sweetened, add *also* 1 tablespoonful of sugar or honey for each avocado. Mix all ingredients thoroughly with the avocado. The mashed pulp may be packed in small rigid containers, although what I usually do is to return the mixture to the scooped-out shell halves, cover the open sides with cellophane, and wrap the joined halves closely with aluminum foil.

Bananas

Oh, yes.

Despite the rhumba-rhythm warnings of Chiquita Banana, I have frozen bananas for use in making ice cream, banana bread (recipe, page 365) and malted milks. I don't advise any woman to do this on a day she feels relaxed and wants to stay that way, because fingers have to fly when you work with bananas—they darken so rapidly.

Slice and mash ripe (but not over-ripe) bananas in a chilled bowl. Mix quickly with acid-treated sugar, 1 cup of sugar mixed with 1½-2 teaspoonfuls of ascorbic-citric acid powder for every three cups of banana pulp. Package in rigid containers, never more to a container than you are sure you will use at one time.

Banana pulp must be defrosted unopened in the refrigerator until it is just soft enough to work with, but not completely thawed. Then it must be used at once in whatever it is you are making with it.

Cooking bananas and plantains:

If you live in an area which receives shipments of tropical plantains, and if you have ever been served a delectable dish of fried bananas in a Latin-American restaurant, you may want to have some in your freezer for your own enjoyment and that of your gourmet friends.

Peel and cut in half lengthwise medium-sized fully ripe plantains or cooking bananas and sprinkle them with lemon juice or with ascorbic-citric acid solution (½ teaspoonful of acid powder for each ounce of water used). Either dredge them in flour or bread crumbs, or dip them in a thin batter of beaten eggs and heavy homogenized cream. Fry them until tender in deep fat, allow them to cool at room temperature, then wrap the halves individually in aluminum foil or cellophane and freeze at once. After they are frozen, several may be put into a large polyethylene bag for easier storage.

Berries, All Kinds

See Chapter 18.

Cantaloupe

See Chapter 19.

Cherries, Sour

Best varieties for freezing: Montmorency, English Morello.

Approximate yield:

1 to 1½ pounds equal 1 pint.

1 bushel (56 pounds) equals from 35 to 45 pints.

This short-season fruit is excellent for pies, purées and juice.

Select fully ripe, flaming red cherries and firm them in ice water for 1 hour. Stem, sort, wash and pit them. If they are to be served without cooking they may be packed in rigid containers and covered with a 50 per cent or 60 per cent syrup if they are not too tart, a 70 per cent syrup if very sour.

Cherries intended for pie are best in a dry sugar pack. Add 1½ cups of sugar for each two quarts of cherries, mixing with a wooden spoon until the fruit is entirely covered and the sugar dissolved.

Pack cherries, either dry or in syrup, in rigid glass or plastic containers that are absolutely liquid-tight. It is best not to use cardboard or paper containers for cherries, as the acid in the juice sometimes never freezes at all, but remains liquid throughout freezer storage.

Cherries, Sweet

Best varieties for freezing: Black Tartarian, Bing, Schmidt, Windsor, Napoleon.

Approximate yield: Same as for sour cherries.

Select richly colored, sweet, fully ripe cherries. Wash them well under cold running water (it is not necessary to firm sweet cherries). Stems may be removed or left on for party and cocktail use. Sweet cherries may be frozen just as they are, or with the pits removed. Needless to say, I do mine the easy way—with stems on and pits in. I like the slightly almond flavor of cherries frozen with their pits intact.

If you have an upright freezer or a fast-freezing shelf, cherries may be frozen just as they come from the tree, spread out on waxed paper until frozen and then packaged loosely in laminated or polyethylene bags. This takes up quite a bit more space than tightly packed cherries in rigid containers, but you may consider the space well spent.

Whole cherries, either pitted or not, may be poured, when washed, directly into containers filled ½ full of cold 40 per cent syrup. To preserve the deep color of red varieties, add 2 teaspoonfuls of ascorbic-citric acid powder to each quart of syrup used. Crumple a piece of cellophane over the top of the cherries before closing the container.

Sweet cherries may also be crushed for use as sundae sauces. Remove pits, chop or crush coarsely (the coarse cutting blade of an ordinary household food chopper does this job efficiently) and put the crushed fruit in a chilled bowl. Add ¾ cup of acid-treated sugar, 1½ teaspoonfuls acid powder for each cup of sugar) to each pound (pint) of crushed cherries. Mix well, and pack at once into rigid plastic or glass containers, leaving ½-inch expansion space. Seal and freeze.

Mixed cherries are very good, if you are able to get both sweet and sour varieties at the time you are in a mood for freezing cherries. Prepare according to the preceding instructions for whole uncooked cherries, and mix them in equal proportions. Pack in 40 per cent syrup or in dry sugar, 1 cup of sugar to each 4 cups of cherries.

Cherry juice has more flavor if both sweet and sour cherries are used, although each may be extracted and frozen alone.

Heat whole pitted cherries very slightly over a low flame to start the juice flowing, and extract the juice through a jelly bag. If very sour cherries are used, add from 1 to 2 cups of sugar per quart of juice (taste it to be sure). If cherries are sweet or mixed, do not add sugar. Allow juice to cool and settle overnight in the refrigerator. Next day, decant the clear juice into rigid plastic or glass containers, leaving 1-inch expansion space. Seal and freeze.

Coconuts

If you live in the semi-tropical belt, you may not think it worth while to freeze coconut meat. But if you live in the colder regions and some inexpensively bought or gift coconuts come your way, you will find frozen coconut meat far superior to the dried shreds bought in packages for use in icings or sprinkled on fruit salads or as one of the exciting ingredients with East Indian curry dishes.

Select coconuts which contain milk (shake them to find out). Split them and grate or shred the meat. Save the coconut milk and use it for moistening the shredded meat. Add ½ cup of sugar to every 4 cups of moistened coconut meat, and pack in bags or containers.

Crabapples

When your tree is loaded with these small, tart apples and you just don't have the time or energy to make jelly, follow the directions for pie apples on page 279 and freeze them temporarily until you are ready for a canning session.

Cranberries

See Chapter 18.

Currants

Red, black or white varieties may be frozen, but the plump bright-red ones are most successful.

Approximate yield: 1 quart (1½ pounds) equals 2 pints.

Select fully ripe currants, wash in cold water and carefully remove all stems and woody parts.

To freeze whole, pack unsweetened, with syrup or with sugar.

Unsweetened: Put into laminated or polyethylene bags, or in rigid containers.

If you have an upright freezer or a fast-freezing shelf, simply spread the currants out on a shelf lined with waxed paper until they are frozen. Then pour them into containers and store them in the freezer.

In syrup: Pack in rigid containers and cover with cold 50 per cent syrup. Leave ½-inch expansion space, crumple a piece of cellophane on top of currants before closing container.

In dry sugar: Put washed and stemmed currants in a chilled bowl and sift over them ¾ cup of sugar for each quart of currants. Stir with a wooden spoon until evenly coated and the sugar is dissolved. Pack in rigid containers, leaving ½-inch expansion space.

Currants may also be crushed and mixed with sugar (1 cup of sugar to each quart of currants) and packed in rigid containers, leaving ½-inch expansion space.

Currant juice may be frozen as a beverage mix or for later jelly-making. Follow the directions for Cherry Juice.

Dates

You may want to freeze the fruits of the date palm instead of putting them away in a cupboard, where they frequently develop an off taste. They are delicious when eaten right from the freezer.

Select fully ripe dates of good flavor and tender meatiness. Wash thoroughly and slit with a sharp knife to remove the pits. Package in laminated or polyethylene bags. (See recipe for Date and Nut bread, page 365.)

Figs

Best varieties for freezing: Kadota, Mission, Calimyrna.

Select fully ripened figs that are soft-ripe and sweet, with tiny seeds and skins which are unsplit but slightly shriveled. Sort out and discard any bruised, sour-tasting or too-soft ones. Wash carefully in ice water to firm them, and remove stems with a sharp knife. Do not handle harshly, as figs bruise easily.

Figs may be frozen whole and unpeeled, whole and peeled, sliced or crushed.

Whole figs, peeled or unpeeled: For breakfast and dessert use, pack in acid-treated 30 per cent syrup. Leave ½-inch expansion space at top of rigid containers.

As a confection, pack dry with acid-treated sugar (1 cup of treated sugar to 5 cups of figs) and package in rigid containers or laminated bags.

Sliced figs for baking may be packed and frozen by either of the above methods.

Crushed figs: Prepare and wash as for whole figs. Crush coarsely with the tines of a steel fork, put through a food chopper or force through a strainer. For every two pints of crushed figs, add ½ cup of acid-treated sugar. Pack in rigid containers, leaving ½-inch expansion space.

Fruit Juices, Citrus

As you discovered soon after the introduction of those wonderful frozen fruit juice concentrates, frozen citrus juices are far superior to the canned varieties. They have a fresher flavor, closely resembling the original fruit.

Select firm, tree-ripened oranges, grapefruit, lemons and limes for juice. Fruit should be heavy in the hand for its size.

Chill the unpeeled fruit in ice water or in the refrigerator before extracting the juice, which should be done with a squeezer or extractor that removes juice without pressing the bitter oils from the rind. Remove seeds. Juice may be strained, but a more nutritious product includes the fine pulp that escapes through the squeezer.

Add ¼ teaspoonful of ascorbic-citric acid powder for each

quart of juice, and mix thoroughly but carefully, so as not to mix in air. Pour into glass (Ball) jars or rigid clear plastic containers, leaving 1-inch expansion space.

Lemon and lime juice to be used in fruit drinks or for flavoring are handily frozen in ice-cube trays. When the cubes of concentrate are solidly frozen, remove them from the trays and pack in laminated or polyethylene bags.

Fruit Juices, Non-citrus

Out of every batch of fruit or berries you prepare for freezing, there will inevitably be some which seem just a speck too ripe for a good product. When these are sorted out, put them in the refrigerator until you have finished the main job, and use them for frozen fresh fruit juices, either as single flavors or mixing several together to achieve a refreshing blend.

Wash and remove stems, pits, seeds and bruised portions. Cook in a stainless steel or aluminum vessel over low heat, simmering until the juice is separated from the pulp. If the fruit is firm and not very juicy, add ½ cup of water for every quart of fruit in the cooking vessel. Do not boil.

When you think all the juice has oozed out that you are going to get, remove from the flame and allow it to cool a little at room temperature. Strain the juice through a cloth jelly bag or cheesecloth and sweeten to taste if desired. Chill overnight in the refrigerator. Add 1 teaspoonful of ascorbic-citric acid powder for each gallon of juice and mix thoroughly. Pour into glass jars or rigid plastic containers, leaving 1-inch expansion space.

Grapefruit

One of my most successful freezer experiments came about when a traveling relative shipped me two large crates of the most delicious grapefruit I had ever tasted. I never had any experience in freezing grapefruit, but it soon became apparent that I was going to get some. Even eating grapefruit at every meal did not diminish that huge pile of grapefruit very fast, and I was growing nervous about the possibility of spoilage.

After cutting the grapefruit in half, I used a handy little gadget I found once in Macy's basement—a grapefruit de-sectioner. It looks like a cleft arrowhead attached to a handle, and it works like a dream. (I must confess to a passion for gadgets. I have drawers full of shrimp de-veiners, oyster shuckers, bean slicers, pea shellers and cherry pitters. And, of course, a garlic press.)

I de-sectioned about three grapefruit at a time, saving the juice in a separate pitcher. For each quart of juice I stirred in ½ cup of sugar and ¼ teaspoonful of ascorbic-citric acid powder. I packed the fruit in 16-ounce plastic containers, each one sufficient for four servings, and covered it with its own treated juice.

Months later, when no one I knew was traveling and the price of grapefruit had gone sky-high in the stores, I enjoyed wonderful-tasting grapefruit sections for breakfast and in salads.

Grapes

My success-story with grapes was purely accidental. A happy mid-week purchase from a pushcart peddler included a large, perfect cluster of unusually beautiful Tokay grapes. I was entertaining the following week-end, and I thought of what a lovely table-decoration this special bunch of grapes would make bedded on ivy leaves. I sighed a little for the fleeting bloom of perfection, realizing that by the time my guests were due the grapes would have been plucked from their stems and eaten. Then I thought of the freezer.

Why wouldn't the cluster remain intact on one of my shelves until just before the party, when it could be put to decorative service? I had done a little experimentation with flower buds and corsages and found it workable. I was afraid to wrap the bunch or put it in a polyethylene bag for fear of spoiling its symmetry, and so I just laid it on a cleared space of a shelf as gently as I would put a baby in its crib.

On the evening of the party, I took the cluster from the freezer and arranged my table decoration. It was gratifyingly beautiful.

One grape dropped off the bunch while I was handling it, and I popped it into my mouth. The next second, and without regrets for a sacrificed centerpiece, I returned the grapes to the freezer. I had never tasted anything quite so delicious as those hard-frozen grapes. The freezing process had made the pulp swell until the skins looked ready to burst from sheer fullness and richness, and the taste was indescribably different.

One of these days I will have strength of character enough to leave frozen grape clusters in the freezer for a long time, taking some out every week or so in true research fashion in order to find out how long they will remain so delectable. So far, however, the longest period I can report is two weeks and three days.

Grapes may also be frozen more conservatively for fruit cocktails, tarts, juice or for use in jelly-making.

Best varieties for freezing: Catawba, Delaware, Ladyfinger, Niagara, Malaga (red or white), Thompson Seedless, Tokay.

Use only firm, ripe grapes with sweet flavor and tender skins, packing the seedless varieties whole and the others in halves with seeds removed. Pack in rigid containers and cover with a 30 per cent or 40 per cent syrup.

Kumquats

Those interesting little orange-colored citrus fruits which appear in the markets during December and January (unless you have a kumquat tree in your southern or western back yard) are nice to freeze as appropriately exotic dessert fruit with Oriental dinners, or for use in sweet fruit breads.

Wash brightly colored, fully ripe kumquats and firm them in ice water. Pack in rigid containers and cover them with 60 per cent or 70 per cent acid-treated syrup.

Lemons and Limes

Peeled slices of lemons or limes for garnishes on fish dishes or to be used in fruit punches or at the cocktail hour may be frozen in light (20 per cent) acid-treated syrup. Use only ¼ teaspoonful of acid powder for each quart of syrup. Pack in glass or rigid plastic containers, leaving ½-inch expansion space.

Mangoes

Residents and visitors in the south are profitably acquainted with these rough-skinned, distinctively-flavored fruits which combine the tastes of pineapple and apricot. They may be frozen successfully for fruit-cocktail or salad combinations, or be there waiting for you when you get an ambitious urge to make chutney for your shrimp curries. After spending long, hot hours in the kitchen making chutney I began to understand why it costs so darned much.

Choose firm, fully ripened mangoes with deep yellow flesh. Wash in cold water, peel, remove pit, cut in ¼-inch slices. Pack in rigid containers or bags without adding syrup or sugar, although they may be sprinkled with a solution of ascorbic-citric acid to retard darkening. Use ¼ teaspoonful of ascorbic-citric acid powder for each ounce of water to make the solution.

Nectarines

This delicate-flavored, smooth-skinned peach variety is not, contrary to popular belief, a horticultural hybrid of peach and plum. As a matter of fact, nectarines are frequently found growing on the same peach trees, often the same branches, as fuzzy-skinned peaches.

Approximate yield:

1 pound equals 1 pint.
1 box (20 pounds) equals from 15 to 20 pints.
1 bushel (48 pounds) equals from 35 to 45 pints.

Choose firm, fully ripened fruit with blushes on their cheeks. Wash in cold water and drain. The fine skin of nectarines need not be peeled, but if the whole fruits are dropped into boiling water for 30 seconds the skins will not toughen in the freezer. Cut nectarines in halves, or slice as peaches.

Nectarine halves: Cover with acid-treated 40 per cent syrup and pack in rigid containers, leaving ½-inch expansion space.

Sliced nectarines: Slice into a chilled bowl and sift over the fruit ½ cup of acid-treated sugar for each quart, lifting the fruit in and out of the sugar with a slotted wooden spoon until the fruit is entirely covered and the sugar is dissolved. Pack in rigid containers or in laminated or polyethylene bags in boxes.

Oranges

Although sweet oranges can be frozen in sections or cut up into small pieces for fruit salads, I have found that they freeze, pack and keep better if they are sliced.

Peel firm, fully tree-ripened oranges and remove the membrane. Cut into ¼-inch slices. Pack in rigid containers and cover with the juice of squeezed oranges to which you have added ½ teaspoonful of ascorbic-citric acid for each quart of juice.

While I absolutely refuse to enter into the California-Florida orange controversy, I will go so far as to mention that California's Valencias and Washington Navels are easy to work with because they are either completely seedless or almost so. Florida Valencias are almost seedless, too. Despite the number of seeds found in the eastern state's varieties, these oranges are likely to be juicy and sweet. California Valencias may require the addition of a little sugar, unless you like your oranges tart.

Peaches

Best varieties for freezing: Elberta, Halehaven, J. H. Hale, Salway, South Haven, Golden Jubilee, Crawford, and yellow-fleshed clingstones.

Approximate yield:

1 to 1½ pounds equal 1 pint.
1 lug (20 pounds) equals 15 to 18 pints.
1 bushel (48 pounds) equals from 30 to 46 pints.

A good number of the wails I hear about home freezing of fruits come from women who have labored long and lovingly with a bushel of peaches only to find, months later, that the frozen product has turned unpleasantly brown. These peaches are adequate for pies, they complain, but they look dreadful when used for shortcake or in fruit salad. What are they doing wrong? Why don't their peaches look like Birds Eye's or even, to a less professional degree, like mine?

For one thing, I disregard the rule which says you must immerse peaches in boiling water to peel them unless I am making purée. It is more difficult to peel and pit peaches swiftly under running cold water, but that is what I do. Before I proceed to the next peach, I slice or halve the first one directly into an icy bath of ascorbic-citric acid solution (2 tablespoonfuls per gallon of ice water). They have a tendency to float, and I thwart them by placing over the peaches a heavy pottery plate to keep them completely covered in the anti-oxidant bath, lifting the plate each time I have another peach to slice in.

When I have about two quarts of peaches all sliced or halved, I lift them out with a slotted wooden spoon and transfer them into containers in the following manner: I spoon half of them into two pint containers which are already a quarter filled with a 40 per cent syrup, adding more syrup to cover the fruit completely, and crumpling a piece of cellophane over the top of the peaches before putting on the lids and sealing them.

The other quart I drain on paper towels, blotting them gently, and return them to a chilled bowl. With a wooden spoon, I turn them over and over in ½ cup of sugar until all the fruit is covered. These peaches I pack dry in rigid containers. Box-and-bag combinations may also be used if you heat-seal the bags.

There is another trick to frozen peaches, however. Even if you have succeeded in keeping them from turning brown in the freezer, they will begin to darken as soon as they are thawed and opened to the air. Therefore—defrost peaches slowly in the refrigerator and serve them while they are still a little frosty. Use them half-thawed for shortcakes. By the time you have put them in shells or on cake and topped them with cream, they will be ready to eat when you put a fork to them.

Peach purée for sauce and for making ice cream or sherbet:

Loosen skins by dropping peaches into boiling water for 1 minute. Put immediately into ice water. Peel enough at one time to make one quart, and remove pits. Crush peaches with the tines of a fork or press through a sieve. Mix ½ to 1 cup of acid-treated sugar (depending on the sweetness of the fruit and the sweetness desired) in the quart of peach pulp and pack in rigid containers, leaving ½-inch expansion space.

Pears

Handle with care, for this fruit also is as temperamental as a prima donna and its pallid flesh will discolor with heart-breaking rapidity if you don't keep one step ahead of oxidation.

Best varieties for freezing: Your geographical location will usually determine what pears are to be found at reasonable prices in local markets or at orchard roadstands. As a general rule, the fine-tasting fresh Bartletts are not completely satisfactory as a frozen product unless cubed for mixing with other fruits in cocktails and served while they are still frosty.

Approximate yield:

1 to 1½ pounds equal 1 pint.

1 bushel (50 pounds) equals from 35 to 45 pints.

Pear halves:

Method 1. Choose firm, ripe fruit without bruises or broken skins and wash them in cold water. Peel, core and cut in halves directly into boiling 30 per cent syrup. If pieces are small, boil for one minute. If they are large, let them remain in the boiling syrup for two minutes. Remove from the flame and allow pears to cool covered by their own syrup.

Remove them from the cooking syrup, drain them in a colander and pack them immediately in rigid containers which are half-filled with *cold* acid-treated 40 per cent syrup. Be sure that pears are well covered with syrup.

Crumple a piece of cellophane over them before closing the lid.

Method 2. Instead of boiling, cut pears directly into a solution of 2 teaspoonfuls of ascorbic-citric acid to 1 gallon of water and dunk them up and down in the mixture until all the fruit is coated. Drain quickly and pack immediately in rigid containers half-filled with acid-treated 40 per cent syrup.

Persimmons

Both the large red Japanese persimmon and the smaller American variety grown in the southern, middle Atlantic and middle western states may be frozen sliced for pies and fruit cocktails, or puréed for puddings and sauces.

Sliced persimmons: Select only evenly rich-colored persimmons that are soft-ripe but not mushy. They should be sweet to the taste, with no hint of the astringent, mouth-puckering characteristic frequently found in insufficiently ripened persimmons.

Wash the fruit and firm it in ice water. Peel and cut into crescent sections, cutting the fruit into a chilled bowl. Sprinkle with ascorbic-citric acid solution (½ teaspoonful per pint of water), lifting the fruit gently with a slotted wooden spoon until you are sure all the pieces are coated. If you wish, you may also sift in ½ cup of sugar for each quart of persimmon slices, coating the pieces in the same manner until the sugar dissolves. Pack in rigid containers or in bag-and-box combinations (heat-seal these).

Persimmon purée: After washing and peeling the fruit, press its pulp through a sieve or food mill. For each quart of persimmon pulp, add ¼ teaspoonful ascorbic-citric acid powder, mixing it well. Pack unsweetened. Or, mix each quart of purée with ½ to 1 cup of acid-treated sugar, depending on taste. Pack in rigid containers, leaving ½-inch expansion space.

Pineapple

Approximate yield: A 2 to 2½ pound pineapple will average about 2 pint containers of sliced, cubed or crushed pineapple.

You don't have to worry a bit about pineapple. It freezes like a charm, and it is mighty nice to have several packages of various cuts on hand for many purposes—baking or broiling with ham roasts or steaks, in salads and fruit cups, as a clean-tasting, refreshing dessert after a heavy meal, or as a topping for cakes and ice cream.

Just be sure when you buy an armload that the pineapples are full ripe and free from decay. If the leaves pull out easily, it is ripe. If the base of the fruit is tender but firm and without any brownish soft spots, it is usually in good condition. Inside, the fruit should be golden-yellow.

The easiest way to peel one of these prickly things for half-round slices is to cut it in half lengthwise, then lay each half fruit-side down on a cutting board and cut into ½-inch slices. Removing the outer rind, core and eyes is then a simple matter. To get rounds, impale the pineapple sideways on a spiked roast board and follow the same procedure. Watch your hands; pineapples bite.

Freeze pineapple rounds or half-rounds unsweetened in rigid containers or trunk-opening lined boxes by placing double thicknesses of cellophane or parchment paper between each slice for easier separation later. They may also be packed tightly into rigid containers and covered with a 20 per cent or 30 per cent syrup. Add to the syrup whatever pineapple juice you have succeeded in catching, or cover it with another fruit juice. Leave ½-inch expansion space.

Freeze pineapple sticks, cubes or wedges by packing tightly, either sweetened in a light syrup and in rigid containers (½-inch expansion space) or just plain, in bag-lined boxes. Crushed pineapple may be mixed with sugar (½ cup per quart) and packed in rigid containers.

Label your pineapple packages carefully as to the style of cut and pack. If you ever plan to use this fruit in gelatine desserts or aspic salads, you must remember to boil it a minute or two before adding it to the gelatine mix, otherwise the mold simply will not jell. (This is because pineapple contains an enzyme which digests protein, and gelatine is protein.)

Plums (and Prunes)

Best varieties for freezing:

Plums: Beauty, Climax, Duarte, Gaviota, Giant, Kelsey, President.

Prunes: Italian, French.

Approximate yield:

1 to 1½ pounds equal 1 pint.

1 box (20 pounds) equals from 14 to 18 pints.

1 bushel (56 pounds) equals from 40 to 55 pints.

While retaining color may not be of primary concern for plums, the use of ascorbic-citric acid will help to retain their flavor.

Select fully tree-ripened fruit of rich color and sweet taste. Firm in ice water before cutting and removing pits.

Pack whole without further ado in laminated or polyethylene bags. Label these packages accurately, for when you are ready to serve them as an uncooked compote, they should be dipped while still frozen in cold water for about 10 seconds before peeling, and then dropped into a cold syrup (30 per cent to 50 per cent) and allowed to thaw in the refrigerator. A delicious dessert for wine-lovers is to put the whole plums, drained of syrup, in a stemmed sherbet glass and cover them with port or angelica. Or, if you prefer dry wines, with claret or sauterne.

Halved plums are better when peeled, pitted and packed in rigid containers covered with syrup (30 per cent to 60 per cent). To this pack add from ½ to 1 teaspoonful of ascorbic-citric acid powder for each pint of syrup used. Leave ½-inch expansion space.

Puréed plums: When you sort your bushel or crate of plums for freezing whole or in pieces, you will undoubtedly find many whose over-ripeness or squashiness make you uncertain about the advisability of freezing them for uncooked desserts. Put these aside in the refrigerator while you are working on the main job, and then prepare them for a very tasty plum purée.

Method 1. Press peeled, pitted soft plums through a sieve or food mill and add ½ teaspoonful ascorbic-citric acid powder per quart, stirring until well mixed. Or, add acid-treated sugar to taste. Package in rigid containers, leaving ½-inch head space.

Method 2. Put whole plums in an aluminum pan and add ¼ cup of water for each quart of fruit. Bring to a slow boil, adding more water if necessary, and cook for two minutes. Turn plums into a bowl and allow them to cool. Pit, then press through sieve or food mill, refrigerate until chilled, and package as above.

Quinces

It may seem a little unusual to freeze quinces, but the frozen preserve is a far superior product to the canned variety. It retains more of the very distinctive quince flavor.

Wash quinces thoroughly under running water, peel and remove the cores. Throw the cores away, but put the scrubbed peelings in a pan and cover them with water. For each quart of water, stir in ½ teaspoonful ascorbic-citric acid powder *or* slice in 1 peeled and seeded orange and 1 peeled and seeded lemon.

Simmer the fruit until tender, then strain the juice and pour it over sliced quinces in another pan. Simmer the quinces in this juice until they are just barely tender. Remove from the flame, add and dissolve 3 cups of sugar for each quart, and allow to cool.

Next, decant the liquid and chill it thoroughly in the refrigerator. This becomes your syrup pack. When the syrup is cold, pour it over the quince slices to cover and pack in rigid containers, leaving ½-inch expansion space.

If your family is fond of quince preserve, try making it this way. There will probably be an empty space under "Q" on your canned-goods shelf from then on.

Rhubarb

Best varieties: Linnaeus, MacDonald, other hothouse stringless types with light pink stalks, small leaves. (Field-grown rhubarb, distinguishable by its rich dark red color and coarse green foliage, is extremely tart. This can be frozen, but requires much more sugar.)

Approximate yield: 1 pound equals 1 pint.

The early spring crop of young, tender rhubarb is best for freezing.

Unsweetened, for pies: Wash under cold running water and trim, then cut into your preferred lengths. Drain thoroughly and pack immediately in rigid containers or in bag-lined boxes. If you keep your stalks long, they may be wrapped and sealed in cellophane, laminated freezer paper or aluminum foil for compact packages. Be sure to label. To make more compact packages, conserving freezer space,

FRUITS FOR FREEZING

DESCRIPTION OF FRUITS FOR FREEZING, AND SEASONS OF YEAR WHEN MOST PLENTIFUL

Fruit	*Description*	*In Season* *
Apples		
Baldwin	Bright red	Nov.-May
Delicious	Medium red, streaked	Oct.-May
Delicious (Golden)	Speckled bright yellow	Oct.-May
Jonathan	Bright red	Sept.-Jan.
McIntosh	Red and green	Sept.-Feb.
Northern Spy	Red, with stripes	Oct.-March
Rome Beauty	Yellow or green, with red markings	Nov.-May
Winesap	Very dark red	Dec.-June
York Imperial	Red and white stripes	Oct.-April
Apricots		
California Blenheim, Tilton, Moorpak, Hemskirt Royal	Rich golden yellow, shading to pink	July-Aug.
Avocados (Alligator pears)	Dark polished green or purplish	Feb.-April, July-Sept.
Cherries (sour)		
Montmorency	Light or medium red	June-Aug.
English Morello	Dark red	July-Aug.
Cherries (sweet)		
Black Tartarian	Dark red, tender, medium size	May-Aug.
Bing	Oxford red, firm, large	June-Aug.
Napoleon	Red and yellow, firm, large	June-Aug.
Schmidt	Almost black, firm, large	July-Sept.
Windsor	Almost black, firm, medium	July-Sept.
Coconuts		Dec.-Jan.
Currants	White, red or black	June-Aug.
Dates	California, Arizona packs	Sept.-Dec.
Figs		
Calimyrna	Large, white or greenish-yellow	June-Aug.
Kadota	Small, white or yellow	June-Aug.
Mission	Large, purplish black	June-Aug.
Grapefruit	Seedless	Oct.-May
Grapes		
Catawba	Oval, red-purple	Sept.-Nov.
Delaware	Small, round, red	Sept.-Nov.
Ladyfinger	Long, light green, large	Sept.-Oct.
Niagara	Large, green	Sept.-Nov.
Malaga	Red or white	July-Nov.
Thompson Seedless	Oval, greenish yellow	July-Feb.
Tokay	Oval, bright red	Sept.-Nov.

DESCRIPTION OF FRUITS FOR FREEZING, AND SEASONS OF YEAR WHEN MOST PLENTIFUL (*Continued*)

Fruit	*Description*	*In Season* *
Kumquats	Oval, orange, small	Dec.-Jan.
Lemons	Yellow or green	All year
Limes	Florida, California	May-Nov.
Mangoes	Rough skin, yellow flesh	June-Sept.
Nectarines	Smooth skin, peach-like flesh	July-Sept.
Oranges		
California		
Valencia	Smooth skin, few seeds, deep color	May-Nov.
Washington Navel	Seedless, deep color, navel	Nov.-May
Florida		
Valencia	Same as California crop	March-June
King	Pale, thick loose skin	Jan.-May
Seed varieties	Yellow to deep orange color	Oct.-Jan.
Peaches		
Freestone		
Elberta	Large, oval, yellow flesh	July-Sept.
Hale	Large, oval, yellow flesh	July-Sept.
Salway	Large, round, reddish flesh	June-Oct.
Crawford	Round or oval, large, yellow flesh	May-Aug., July-Oct.
Clingstone	Yellow-fleshed varieties are best for freezing	June-Sept.
Pears		
Bartlett (summer)	Bell-shaped, smooth red and yellow skin. Best frozen in cubes for salads.	Sept.-Oct.
Bosc	Dark, mottled yellow, long-necked	July-March
Comice	Green-yellow blushed with red, large	Oct.-Dec.
D'Anjou	Yellow touched with red, large	Oct.-April
Forelle	Blushed green, speckled	Oct.-Nov.
Hardy	Green, russeted skin, off-center stem	Aug.-Nov.
Kieffer	Coarse flesh (steam blanch)	Oct.-April
Seckel	Bronze, red-cheeked, small egg shape	Aug.-Jan.
Persimmons		
Japanese (grown in California, Texas)	Large, reddish	Oct.-Feb.
American (grown in south, middle Atlantic and middle western states)	Small, golden-brown	Oct.-Dec.

DESCRIPTION OF FRUITS FOR FREEZING, AND SEASONS OF YEAR WHEN MOST PLENTIFUL (*Continued*)

Fruit	*Description*	*In Season* *
Pineapples	Imported from Hawaii, Cuba, Puerto Rico	March-Aug.
Plums		
Beauty	Heart-shaped, crimson; streaked flesh	July-Sept.
Climax	Dark brown-red, large; yellow flesh	June-Oct.
Duarte	Dotted dark red, large; red flesh	June-Sept.
Gaviota	Large, cone-shaped, yellow to red, small pit; yellow flesh (excellent)	June-Oct.
Giant	Large, egg-shaped, purplish, yellow flesh	June-Oct.
Kelsey	Very large, round, reddish yellow	June-Oct.
President	Very large, oval, purple	June-Oct.
Prunes		
Italian	Deep blue skin, tart flavor	Aug.-Nov.
French	Small blue oval, sweet flavor	Aug.-Nov.
Quinces		Sept.-Nov.
Rhubarb		
Linnaeus, MacDonald	Light pink, small leaves, mild flavor, almost stringless	March-Aug.
Field-grown	Dark red, coarse green foliage, tart flavor	May-Aug.

* Seasonal information compiled from statistics furnished by *Chain Store Age*, a grocery trade publication. This is a national average. For the peak seasons in your own region, consult your county agent.

blanch the stalks in water for 1 minute. This makes them slightly limp and easier to pack closely.

Sweetened, for dessert: Wash and cut rhubarb and cook it exactly as you do for the table, omitting sugar. When it is cool, pack it in rigid containers and cover with 40 per cent syrup. Leave ½-inch expansion space.

Sweetened purée, for sauce: Boil (1 cup of water to 2 quarts of cut rhubarb) for 2 minutes and let it cool. Press through sieve or food mill. Add ½ to ⅔ cup of sugar for each quart of purée and pack into rigid containers, leaving ½-inch expansion space.

Length of Storage Time for Fruits

When carefully selected, properly prepared and packed in moisture-vapor-proof containers with the added protection of ascorbic-citric acid for those varieties suggested, most fruits, purées and juices can be stored at 0°F. up to a year; at 5°F., up to 8 months. Pre-cooked fruits should be used before 4 months have elapsed.

18. Freezing Berries

Berries of all types are just about the easiest of foods to freeze, for they seem to have a magical ability to retain their rich color and fresh flavor for a long time without being in the least fussy about added preservatives such as anti-oxidants. They have apparently spent their brief but beautiful lifetimes busily amassing quantities of ascorbic acid (Vitamin C) while hoarding a natural sugar content which makes them irresistible to birds and people.

It is not necessary to add to berries the ascorbic-citric acid preparations recommended for most fruits, and you may sweeten them or not as family tastes dictate.

If you pick or buy berries when they are at their ripe, colorful best, handle them tenderly, wash them well, pack them in moisture-vapor-proof containers according to the simple directions given for each variety and use . . . you will capture for mid-winter enjoyment all the nostalgic, memorable flavor of a summer day.

To most people, especially city dwellers, berries mean raspberries, strawberries and blueberries snatched eagerly from a fruit store or roadside stand during their fleeting market season. But there are enough berries to make a long poem, and a good many of them grow wild on bushes not very far from your house by car and pail.

A picnic in the country on a summer day, wearing long sleeves, jeans or old slacks and a sun-shading hat, is triple fun for the whole family. It's fun to find and pick berries, fun to pack them away in the freezer, and fun to have fresh-berry shortcake on the table when it's snowing outside.

Note to owners of upright freezer plate cabinets or chests with fast-freeze compartments: All berries for pies and other cooked desserts may be frozen unsweetened by just turning them out in a single layer on a piece of waxed paper placed on the freezer plate as soon as they are sorted, washed and drained. They will freeze with astonishing rapidity, after which they can be poured like buckshot into bags or rigid containers and stored away.

Blackberries

Best varieties for freezing: Alfred, Blowers, Eldorado.

Approximate yield:

1 to 1½ pints equal 1 pint.

1 crate (24 pints) equals from 30 to 36 pints.

Select fully ripe but not over-mature berries that are dark and glossy, discarding those which are seedy, bruised or poorly colored, with some green areas. If possible, go berry-picking in field or market early in the morning and plan to freeze your haul the same day.

After sorting and removing any leaves or stems, put black-berries in a colander and wash carefully in ice water, never more than a quart at a time, moving the colander gently up and down in the water to reach all berries without favoritism. Turn the berries out on absorbent paper towels and drain them almost dry.

Blackberries for uncooked desserts may be packed in 30 per cent to 50 per cent syrup (see directions for making Syrup on page 280) or tumbled in sugar (1 cup of sugar or less to 5 cups of berries). Sift sugar on to berries in a chilled bowl and lift them gently in and out with a wooden spoon until the sugar is dissolved. In either case, pack berries in rigid containers, leaving ½-inch expansion space. Crumple cellophane over top of syrup-packed berries before closing lid. Seal and freeze.

Blackberries for cooked desserts, such as pies, cobblers or preserves, may be packed directly into containers as soon as they are sorted, washed and drained, without the addition of sugar or syrup. Leave ½-inch expansion space.

Crushed berries for sauces or frozen desserts: Soft-ripe blackberries sorted out from your regular pack may be washed and drained separately, then crushed or pressed through a sieve. Add ½ cup of sugar to each pint of crushed berries and stir until dissolved. Pack in rigid containers, leaving ½-inch expansion space.

Black Raspberries

See Raspberries.

Blueberries

Unless you have an upright freezer or a fast-freeze compartment (see Note on page 313) it is best to crush these berries slightly before freezing them with or without sugar or syrup, for the skins have a tendency to toughen. Another method to prevent toughening is to subject them to brief steam blanching, which seems an unkind thing to do to blueberries except those intended for cooking.

Best varieties for freezing:

Cultivated: Concord, Harbur, Jersey, Pioneer, Rancocas, Rubel, Stanley.

Wild: high or low bush varieties are equally good, if the berries are large and sweet enough to bother with.

Approximate yield:

1 to 1½ pints equal 1 pint.

1 crate (24 pints) equals from 30 to 36 pints.

Remove leaves and stems and sort carefully, removing all berries that are puny in size or under-ripe. Wash in ice water and drain thoroughly.

For pies, muffins and cooked desserts, pack dry and unsweetened in rigid containers or laminated or polyethylene bags.

For uncooked desserts, pack in cold 40 per cent syrup in rigid containers, leaving ½-inch expansion space. Or crush slightly and use a dry sugar pack, ½ to ¾ cup of sugar to 4 cups of berries, tumbling berries in bowl with a wooden spoon until sugar is dissolved.

If skins seem to be somewhat tough while the berries are still unfrozen, hold them over steam for 1 minute. When they are cool, add sugar or cold syrup and proceed with packaging.

Boysenberries

Best varieties for freezing: Giant, Hybrid, Phenomenal.

Approximate yield: Same as Blackberries.

This delectable maroon berry which looks something like a large blackberry which forgot to get black is the offspring of the polygamous marriage of loganberry, raspberry and blackberry, and is named for Rudolph Boysen, the botanist who first produced it.

Follow directions given for Blackberries.

Cranberries

Why limit your enjoyment of fresh cranberry sauce to the holidays, when this beautiful, distinctive berry is traditional? You have a freezer, so tradition be blowed! Chicken or pork tastes just as good in May as December, and the flavor of both is enhanced with cranberries in one form or another whatever the season.

Best varieties for freezing: Early Black, Searles Jumbos, McFarlins, Late Jerseys, Late Howes.

Approximate yield:

1 pound equals 2 pints.

1 peck (8 pounds) equals 14 to 16 pints.

1 box (25 pounds) equals 45 to 50 pints.

Cranberries are as hardy as they are beautiful and may be prepared and packed for freezing in almost any way your heart and family desire. I usually just freeze mine on a plate of my upright freezer and scoop them into polyethylene bags. Once in a while, I also wheedle my mother into whipping up a batch of what I call "Carlotta's Careless Cranberries," which are just raw cranberries put through the food chopper along with whole oranges (skins, pulp and all) and mixed with not very much sugar and, I'm sure, a dash of magic. I have been accused, justly, of hiding the last pint of this incomparable relish behind a large roast in the freezer until the next gallon was in production.

Cranberries should be firm, full-colored and glossy, with no mealiness. Stem and sort them, tossing out any defective, shriveled or soft berries. Wash in cold water and drain.

Pack dry and unsweetened in laminated or polyethylene bags or in rigid containers. They may also be packed whole in containers and covered with cold 50 per cent syrup.

Cranberry purée is prepared from the berries after they have been sorted and washed. Cook in water (2 cups of water per quart of berries) until the skins pop, then press through a sieve or food mill. Add sugar to taste, from 1 to 2 cups for each quart of purée, and allow it to cool. Refrigerate, then pack into rigid containers, leaving ½-inch expansion space.

Dewberries

These are really a species of blackberry, somewhat smaller and juicier than their prototype.

Best varieties for freezing: Lucretia, Young's.

Approximate yield: Same as Blackberries.

Follow directions given for Blackberries for freezing.

Elderberries

I practically never see these in markets. I am not even sure that they are cultivated by growers except possibly for wine-making. Perhaps wild elderberry shrubs and trees by the side of country roads are watched as jealously and impatiently nowadays by little girls and boys and their parents who, like my family of quite a few years ago, shouted, "They're ready!" when the graceful clusters of bead-like deep red berries popped juicily between exploring fingers.

The scramble, the insect bites and thorn scratches, the hours under a hot sun were often rewarded with no more than a medium-sized pailful; for elderberries are very small, just the right size for a bird's eager gulp. The birds loved them as much as we did, and usually got up earlier. But the pailful made a pie—and what a pie! If we had managed to find a particularly prolific patch, there were elderberry pancakes for breakfast the next day.

Someday I am going to dig dozens of the wild elderberries I saw growing along a nearby road, even if I have to risk poison ivy to get at them. And I shall freeze, at the very least, 52 quarts of elderberries.

They will be washed, stemmed, frozen quickly on a freezer plate and poured like pearls into 52 quart-size polyethylene bags, and I shall make one fat elderberry pie every Saturday for a year.

Gooseberries

These odd, melon-striped green berries have always reminded me of the striped glass marbles that were so highly prized in schoolyard tournaments. They make very interesting tarts and pies.

Approximate yield:

1 to 1½ pints equal 1 pint.

1 crate (24 quarts) equals from 32 to 36 pints.

If gooseberries are being frozen for tarts and pies, get them when they are fully mature and pack them unsweetened in rigid containers without sugar or syrup. For later jelly-making, gooseberries may be frozen while slightly under-ripe, packed in rigid containers and covered with 50 per cent syrup. Leave ½-inch expansion space for either pack.

Huckleberries

Follow directions as given for Blueberries.

Lingberries (Lingonberries)

Follow all directions as given for Cranberries, which, in a miniature way, lingberries resemble.

Loganberries

A species of blackberry, with some of the characteristics of raspberries as well. Follow directions for preparing and freezing Blackberries.

Raspberries

Red, black and purple varieties of raspberry may all be frozen in a number of ways for a number of delicious desserts. It is best to reserve berries that abound in seeds for purées or juice, freezing whole the less seedy cultivated varieties.

Best varieties for freezing:

Red: Latham, Indian Summer, Cuthbert, June, King.
Black: Cumberland, Bristol, New Logan.
Purple: Sodus.

Approximate yield (with luck):

1 pint equals 1 pint.
1 crate (12 quarts) equals from 20 to 24 pints.

Sort out under-ripe or seedy berries, discarding the former and saving the latter for purées or juice. Wash carefully in ice water, placing only about a quart at a time in the colander. Drain on absorbent paper towels.

Freeze in direct contact with upright plate or fast-freeze compartment, if you have one, then simply pour unsweetened frozen berries into polyethylene or laminated bags.

Other methods:

1. Pack in laminated bag-and-box combinations, seal and freeze.

2. *Sugar pack:* Add ½ to ⅔ cup of sugar to each quart of berries and tumble them carefully with a wooden spoon until all berries are covered and all sugar is dissolved. Pack in rigid containers, leaving ½-inch expansion space.

3. *Syrup pack:* Put washed berries in rigid containers and cover with cold 30 per cent or 40 per cent syrup, leaving ½-inch expansion space.

Raspberry purée, wonderful as sauce for ice cream or cake, or as the base for summer sodas and other fruit drinks, is simply made by pressing prepared berries through a sieve or food mill. Add ½ to 1 cup of sugar to each quart of purée, depending on the sweetness desired. Mix thoroughly until sugar is dissolved, then pack in rigid containers, leaving ½-inch expansion space.

Raspberry juice: After berries are washed, crush them slightly and heat them just a little over a low flame in order to start the flow of juice. Strain through a jelly bag and sweeten the juice, if desired, with ½ to 1 cup of sugar to each quart of juice. Stir until sugar is dissolved, and allow it to cool.

If juice appears cloudy after it is strained, let it stand overnight, covered, in the refrigerator. Pour off the clear juice in the morning, and add the sugar at this time. Pack in glass jars or plastic containers, leaving ½-inch expansion space.

Strawberries

These are the queens of the freezer, as anyone knows who cheerfully passes up boxes of them marked 75 cents in November because she has pints and pints of them at home that were purchased and frozen in June, when they were 25 cents a box.

Best varieties for freezing: Whatever you can find, just so long as the berry is of fair to large size, of uniform bright color to the very center and not too sour.

Among the excellent freezing species are: Sparkle, Dorsett, Fairfax, Premier, Julymorn, Gem, Senator Dunlap, Big Joe, Catskill, Tennessee, Chesapeake, Fairpeake, Red Star, Temple, Mastodon and Shasta. (Many crops on the market are known only by the name of the state which grows and ships them.)

Approximate yield:

2 quarts equal 3 pints.

1 crate (24 quarts) equals from 34 to 36 pints.

You have noticed that the frozen strawberries you buy under various trade names in stores are usually either sliced or slightly crushed in their own juice. There is a very good reason for this. While strawberries may be frozen whole, either in a dry sugar or a syrup pack, they may be disappointing. Unless you feel reckless and can afford to because you have simply gallons of strawberries to play with, I suggest that you avoid vexation with this most delicious of all berries by cutting them in half, slicing them or crushing them just a little before packaging.

If you are going to freeze them whole in spite of anything I say (and why shouldn't you?) the best way, of course, is to fast-freeze them *without washing or hulling* by arranging them in a single layer on wax paper on an upright contact plate or in the fast-freeze compartment. Then pack them in laminated or polyethylene bags or in rigid containers for storage. When you are ready to use them, remove them from the freezer, wash under cold water, remove caps (or retain them for dunking handles), and serve at once while they are still frosty. This method is recommended for short-term storage only.

No upright contact shelf or fast-freeze compartment? You can still freeze them whole in a syrup or sugar pack, preferably in rigid containers.

For any type of pack, select richly colored fruit without green or yellow areas inside or out. Berries should be mature but firm. As you sort, save the softest ones for purée or juice, holding them under refrigeration until you are ready for them.

Whole (with reservations); sliced or barely crushed strawberries: Wash berries in ice water, drain thoroughly on absorbent paper towels, and remove hulls with a huller, sharp knife or nimble fingertips (never squeeze hulls off).

Syrup pack: Partially fill rigid containers with 40 per cent or 50 per cent cold syrup, add berries, cover with more cold syrup to within ½ inch of top. Crumple cellophane over top of berries before closing lid.

Sugar pack: Turn berries into bowl and sift over them ½ to ¾ cup of sugar for each quart of fruit. Tumble berries gently in sugar with wooden spoon until fruit is entirely coated and sugar is dissolved. Pack in rigid containers, leaving ½-inch expansion space.

Strawberry purée: Wash, sort and hull berries and press them through a sieve or food mill. Add ½ to ¾ cup of sugar for each quart of purée, mix well, and pack in rigid containers, leaving ½-inch expansion space.

Strawberry juice: Prepare as for purée, but strain juice through a jelly bag. Leave unsweetened, or add ½ to 1 cup of sugar for each quart of juice and pour into glass or rigid plastic containers, leaving ½-inch expansion space.

Youngberries

A cross between the blackberry and the dewberry, this species is handled the same way as Blackberries.

ALL BERRIES. Seal and freeze each package immediately. Do not wait until you have them all done before putting them in the coldest part of your freezer.

When to Buy Berries

Except for strawberries, which are in best supply during May and June, and cranberries, in prime condition between September and December, most of the berries described reach their peak season in June and July.

19. Freezing Melons

There are more melons to meet the speculative eye of a freezer owner than just cantaloupes and watermelons, and they may all be frozen for triumphant, economical serving in the depth of winter, when, if they are available at all, they are worth their weight in gold.

Inasmuch as the same procedure is used for each variety, general directions will be given for all. Refer to the melon chart on page 324 for descriptions, tests for ripeness and best purchasing seasons.

Approximate yield: From 1 to 1½ pounds of any melon will yield 1 pint of melon balls, cubes or slices in syrup.

Procedure: Cut melon in half (watermelon in quarters) and remove seeds. If you intend to freeze slices or cubes, peel the melon and slice it with a sharp stainless steel knife into strips no larger than ½ × ½ × 2 inches, or into ¾-inch cubes. For balling, leave halves or quarters unpeeled and

FREEZING MELONS

DESCRIPTION OF MELONS FOR FREEZING, AND SEASONS OF YEAR WHEN MOST PLENTIFUL

Melon	*Description*	*Test for Ripeness*	*Peak Season*
Cantaloupes			
Hale's Best and other rosy-fleshed varieties	Coarse, green-yellow rind with well-developed grayish netting	Slightly sunken stem end; slightly resilient blossom end	May-Aug.
Rocky Ford and other green-fleshed varieties	Same	Same	June-Sept.
Casabas, creamy-white flesh	Large oval or round, wrinkled yellow or greenish skin	Somewhat soft blossom end	July-Nov.
Cranshaws, salmon-colored flesh	Hard gold-green rind, smooth and faintly ribbed. Round at base, tapered stem end	Yellowing skin, fairly soft blossom end, pungent aroma	Oct.-Nov.
Honeyballs	Cross between cantaloupe and honeydew; netted skin, pale flesh	Same as for cantaloupes	June-Sept.
Honeydew, pale green flesh	Smooth, creamy white or pale green skin; oval shape	Slightly soft and resilient blossom end	June-Nov.
Persians, pink flesh	Resembles large round cantaloupe, with heavily netted yellow or greenish skin	Fairlv soft blossom end	June-Oct.
Watermelons			
Cannon Ball	Ball-shaped, with solid green rind	Plug	May-Sept.
Cuban Queen	Long oval, pale green	Plug	April-Sept.
Dixie Belle	Striped dark and light green rind; large oval or roundish shape	Plug	May-Sept.
Tom Watson	Long oval, solid dark green rind	Plug	May-Sept.

NOTE: Slapping or thumping a watermelon is not a revealing test for ripeness, for an over-ripe and consequently mealy melon will often give out the same dull, hollow thud as a prime ripe one.

Many farming areas are experimenting with the development of seedless or almost seedless varieties. When found, these are excellent for freezing purposes.

scoop out the flesh with the small end of a potato-baller.

Pack fruit in rigid containers and cover with cold 30 per cent syrup to within ½ inch of top.

A tastier product may be achieved by adding 1 or 2 level teaspoonfuls of ascorbic-citric acid powder for each quart of syrup used. Interesting variations in flavor can be effected by using fresh orange juice, pineapple juice or ginger ale instead of syrup.

Crumple a piece of cellophane and put it on top of the syrup-covered melon before closing, sealing and freezing the containers.

For color and menu variety, different melon balls or cubes can be combined, i.e., watermelon and honeydew balls; Cranshaw and Persian cubes for salads and fruit cocktails.

Do not allow melons to thaw completely. Place unopened container in the refrigerator and serve the melon while it is still a little frosty.

20. Frozen Food Cookery

When you have bought a quantity of fresh food at money-saving prices, prepared it carefully, packaged it protectively and frozen it rapidly for short- or long-term storage, you have gone only halfway on the road to realization of your freezer's contribution to better living. The proof of more than puddings is in the eating. Your freezer has done its job of efficient preservation; now it is up to you to make the most of it.

Often and sadly I have watched a woman take from her freezer a thick sirloin steak or standing rib roast and cook it to an undeserved death in a hot oven because she did not know how to time it. I have watched other homemakers ruin the flavor and nutritive value of frozen vegetables by boiling them mercilessly in too much water and then, after the water has received all the flavor and vitamins, *throw it down the drain.* I have heard ordinarily good cooks complain about the texture and taste of fried frozen chicken, or of frozen fried chicken. I have been told that Mrs. So-and-So will never again freeze strawberries because they are tasteless.

In the far-reaching research necessary for the writing of this book, I came across a wonderful word, *organoleptic,* defined in Webster's Unabridged Dictionary as "(a) affecting or making an impression upon an organ or the whole organism; (b) capable of receiving an impression." For example, organoleptic tests are made by giving a jury of tasters a certain food to classify according to its flavor. Out of a jury of ten, perhaps four will find the food utterly delicious; three will shruggingly pass it as "O.K., but not outstanding"; two may mark down, "indifferent," and the tenth may spit it out in disgust.

There is an organoleptic influence at the basis of all cooking, of course. There are few universal or absolute standards to guide the cook; she usually prepares and serves food the way her own family likes it. A well-done steak may make you and me wince, but that is the only way Mrs. Smith's husband and oldest girl will eat it. For some discriminating palates, no recipe can have too much garlic in it, while others refuse to eat a salad whose bowl was barely kissed by a cut clove. I won't eat lamb, and my best friends think I'm crazy.

This brief excursion into the psychological subtleties of individualistic food evaluations is intended to explain why

I am reluctant to dictate inflexible rules for the cooking of frozen foods.

Reluctant though I am, however, I must in complete candor say that my own long-time and varied experiences have taught me a few simple truths about the cooking or serving of foods taken from my freezer, and I shall pass these along to you for what they are worth before setting down alternative methods which may be used at your discretion according to the circumstances under which a meal is being prepared. For example, although I have found complete slow defrosting of all beef cuts to give me a better cooked product, there have been times when I was faced with the necessity of preparing a steak dinner in a hurry. At those times, I unhesitatingly broiled the steaks from their frozen state—and without apology.

Things I have learned:

Fruit, berries or melons are at their best when served before they are entirely defrosted.

Vegetables should never be thawed but should be dropped into a minimum of rapidly boiling salted water and cooked briefly. You can, however, either half-thaw spinach—or break the frozen block apart with a fork as it heats.

Meats, especially beef roasts and steaks to be served rare or medium rare, are best defrosted slowly in their wrappings by putting them in the refrigerator the night before they are going to be cooked.

Veal and pork roasts, because of their longer cooking times, may be cooked from the frozen state.

Stews and pot roasts are often better when started cooking from the frozen state, as the meat juices blend with the liquid used to add flavor and food value.

Roasting chickens, turkeys and other poultry are juicier and more tender if they are defrosted slowly in the refrigerator and put into a moderately slow oven as soon as they are thawed.

Cut-up chickens for broiling or frying are also better when allowed to thaw slowly. If you wish, the packages may be opened and the pieces loosely re-wrapped separately to speed the defrosting.

Fricassee or soup chickens should be put from the freezer directly into the pot.

Fish and seafood are best when defrosted in the refrigerator and cooked as soon as the pieces can be separated.

Pre-cooked stews, casseroles, etc., should be heated slowly from the frozen state in the top of a double boiler or in the oven. If leftovers have been frozen and will not stand much more cooking, allow them to thaw only partially before heating and serving quickly.

HOW TO USE FROZEN FOODS

Fruits, Berries and Melons from the Freezer

To serve uncooked in cocktails, salads and desserts, best results are obtained by leaving the unopened containers in the refrigerator to thaw. Faster defrosting is achieved at the expense of some flavor by letting unopened containers stand at room temperature, or by placing them in bowls of cool water.

Thawing Time of Frozen Fruits (Per Pint Container)

On refrigerator shelf: 5 to 8 hours, depending on degree of cold within the refrigerator.

At room temperature: 2 to 4 hours, depending on degree of cold or heat in room.

In bowl of cool water: 30 minutes to 1 hour.

To cook frozen fruits, thaw by any of the methods given on page 325 until the pieces can be separated (if you are making a pie) or simply empty the frozen contents into a saucepan. Proceed as if you were using fresh fruit, keeping in mind the amount of sugar or syrup used previously when you prepared it for the freezer.

Remember, too, that frozen fruits are likely to have more juice than your recipes demand. To avoid leaky pies or too-moist cakes when using frozen fruits, either add more thickening or drain off the excess juice.

You will be glad of the care and thoroughness you took in labeling packages for the freezer, for thawing time is influenced by the method you used in packing.

Unsweetened fruits or berries take longer to thaw than those packed in either dry sugar or syrup; and fruit covered with syrup takes longer than fruit mixed with dry sugar.

Remove from the freezer only as much fruit as you will serve, for it will lose quality and flavor if allowed to stand for any length of time after it has thawed. If it is impossible to avoid leftover thawed fruit, better cook it or use it in a baked recipe as soon as you can to prevent waste. Cooked, it will keep for several days in your refrigerator. Baked, it can be refrozen if necessary as part of cake, muffins or sweet bread. Tarts and pies, however, are best if the fruit or berries have not been permitted to stand after thawing.

Vegetables from the Freezer

You will notice that most of the commercially frozen and packed vegetables you buy at the store include printed instructions for cooking on their labels. There is good reason for this. The food companies learned that most of the early complaints about the taste of frozen vegetables came from women who overcooked these foods and thereby sacrified not only flavor, but also color, texture and vitamins. Always remember that the time you spent in blanching vegetables before you froze them *is part of the total cooking time.*

Special Cooking Instructions for Corn on the Cob. (Of the following methods, I prefer the first.)

Method 1: Put frozen, unwrapped corn in enough cold water to cover it completely and set over high heat. When water comes to a fast boil, lower heat and simmer for no more than one or two minutes.

Method 2: Thaw corn in its wrapper until all ice crystals have disappeared. The best method is to allow it to thaw slowly in the refrigerator, although it may also be defrosted at room temperature. Have a large kettle of water boiling furiously on the stove. When the corn is completely thawed, drop ears into the boiling water and cook rapidly until tender, *rarely more than 5 minutes and frequently less.* Do not overcook, or the corn will lose its flavor and become watery.

Method 3: Roasting: Preheat oven to 350° and roast corn from the frozen state for 20 minutes. Or wrap frozen ears in foil and roast them on an outside barbecue grill or in your fireplace, turning frequently, until kernels are piping hot. Best handle these with tongs.

Method 4: In pressure cooker: (a) Place frozen corn in pressure pan with ½ cup of hot water. When timing begins, pressure-cook for no more than 2½ or 3 minutes. Cool quickly as recommended for your type of pressure pan.

(b) Excellent results have been reported to me by a middle-western lady who uses half milk, half water in her pressure pan when cooking corn on the cob. I have not tried it, but it sounds reasonable.

TIMETABLE FOR COOKING FROZEN VEGETABLES

Vegetable	*Cups of Water Per Pint Package*	*Cooking Time (Minutes) After Water Returns to Boil**
Artichokes	1	15-30 (according to size)
Asparagus cuts	¼ to ½	4-8
Asparagus spears	½	6-12
Beans		
Baby lima	1	6-10
Large lima	1	15-20
Green or wax		
Pieces	½	12-18
Frenched	½	5-10
Beets		
Whole cooked	½	Until heated.
Cooked slices or cubes	¼	Until heated.
Beet greens	½	6-12
Broccoli	½	4-8
Brussels sprouts	½	4-9
Cabbage	½	5-10
Carrots	½	4-8
Cauliflower	½	4-8
Celery	½	4-6
Collard greens	½	6-12
Corn on cob	See special instructions on page 331.	
Corn		
Kernels	½	3-6
Cream style	¼ c. milk, if needed	Heat, do not boil, until tender.
Eggplant	Use in recipes. If pre-cooked, heat in oven.	
Kale	½	8-12
Kohlrabi	½	8-10
Mushrooms	Add to recipes, or sauté in butter.	
Mustard greens	½	8-12
Okra	½	10-20
Onions	Add to recipes.	
Parsley	Add to recipes.	
Parsnips	½	10-15
Peas	½	5-8
Peppers	Used blanched halves for baking without further cooking; fry pieces or add to recipes.	

* Lower heat when water returns to boil.

TIMETABLE FOR COOKING FROZEN VEGETABLES
(Continued)

Vegetable	*Cups of Water Per Pint Package*	*Cooking Time (Minutes) After Water Returns to Boil*
Potatoes		
Sweet	Heat through by baking, frying or glazing in oven.	
White		
Baked and stuffed	Heat in oven.	
French-fried	Heat through on cookie tin or sheet of aluminum foil in oven.	
Pumpkin	Thaw, use for pies; heat in oven or double boiler if served as vegetable.	
Soybeans	½	8-12
Spinach	½	3-6
Squash		
Summer	¼	8-12
Winter	Heat pre-baked halves in oven; heat cooked mashed squash in double boiler.	
Succotash	¼	10-15, add milk and continue cooking over low heat (do not boil) until tender.
Swiss chard	½	8-10
Tomatoes	Heat over low flame without adding water.	
Turnips	½	8-12
Turnip greens	½	8-10

NOTE: Vegetables are started cooking while they are still a solidly frozen block. Usually by the time the water in which you place them has returned to a boil, the vegetables can be broken apart with a fork to assure even cooking. Do not discard the water in which vegetables have been cooked. Use it as the basis for sauces, or add it to soups and stocks.

TIMETABLE FOR COOKING FROZEN VEGETABLES IN A PRESSURE PAN

Vegetable	*Cups of Water Per Pint Package*	*Cooking Time (Minutes) When Pressure Starts*
Artichokes	1/4	7-10 (according to size)
Asparagus cuts	1/3	1/2
Asparagus spears	1/3	1
Beans		
Baby lima	1/2	3/4
Large lima	1/2	2
Green or wax	1/3	3/4-1
Frenched	1/4	1/2
Beet greens	1/3	3/4-1
Broccoli	1/3	3/4
Brussels sprouts	1/3	1
Cabbage	1/3	1/2-1
Carrots		
Sliced	1/4	1/2
Whole	1/3	1
Cauliflower	1/3	1/2
Celery	1/4	1/2
Collard greens	1/3	3/4
Kale	1/3	3/4
Kohlrabi	1/3	3/4
Mustard greens	1/3	3/4
Okra	1/4	1
Parsnips	1/3	3-5
Peas	1/3	1/4
Soybeans	1/3	1
Spinach	1/4	3/4
Squash, summer	1/4	1
Swiss chard	1/3	3/4
Tomatoes	None	1/2
Turnips	1/3	2
Turnip greens	1/3	3/4

NOTE: The times given in both tables are for vegetables which are cooked at sea level or not much above it. Higher altitudes require slightly longer cooking.

Meat from the Freezer

To thaw or not to thaw?

That is the question asked most frequently by new freezer owners and even by veterans who are not quite satisfied with the results of their customary methods.

Replying to that question is not simple. There is no one answer which applies to all meats or to all cooks. Generally speaking, and most manuals on home freezing do speak generally, meat *can* be cooked either from the frozen or the thawed state, and each method has supporters.

Nevertheless, there are certain truths which are self-evident. For example, it is self-evident that you cannot shape hamburger patties from a pound of ground meat which is frozen solidly in a block. You had forgotten that you used up your pre-shaped patties, and you just came home to whip up a fast meal. But by the time that lump of meat defrosts in the refrigerator, which you know is the best place for it to defrost, the dinner hour will have passed into oblivion—and so, as a matter of fact, will tonight's late movie on television.

Even at room temperature, a pound of ground meat takes at least four hours to defrost to the point where it can be molded. In front of an electric fan it may take only an hour, but suppose you haven't got a fan? Beyond swearing in an unladylike manner and vowing never again to underestimate your need for keeping thousands of hamburger patties in the freezer, what can you do?

You can, just this once, remove the ground meat from its wrapping, re-wrap it securely in aluminum foil and put it in a bowl of cool water, adding warmer water gradually until, testing, the outer surfaces of the meat show signs of softening. Unwrapping the meat again, you can with a heavy serrated knife or small freezer saw separate the meat into smaller lumps and then, with clean, warm hands, work the lumps until the meat bends to your will and can be fashioned into seasoned patties. This is all dreadfully unorthodox, but emergencies demand drastic measures. (I am assuming, of course, that you had wrapped the hamburger in meal-size packages. *Don't* refreeze half-thawed meat! See page 88.)

When your husband comes home and hungrily sniffs the pan-broiling hamburgers, he will probably say, "What a cinch! You just open the freezer and pop the stuff in a pan and dinner's ready." If he does, it is wise to turn away and pretend to be looking at something else. He would not understand the expression on your face.

Having cooked every possible kind of frozen meat in every possible way, I will stray far enough away from my unwillingness to "make rules" to say that by far the most satisfactory method I know is to thaw meat—all meat—completely and slowly. This necessitates knowing tonight what you are going to have for dinner tomorrow, as the timetable for thawing will reveal.

The reason why thawed meat provides a better cooked product is fourfold: Thawed meat cooks more evenly. It uses less fuel. It shrinks less (provided you use the low cooking temperatures recommended in the methods tables, pages 334-38. It stays juicier.

Meat should be thawed in its wrappings and opened over a bowl or pan in order to catch any drippings. Add these drippings to gravies or stock, as they are full of valuable B vitamins.

All meat should be started cooking as soon as it is thawed.

Explanation of Cooking Methods

Meat has a dual personality. From the standpoint of nutritive value, it makes comparatively little difference what animal or what part of that animal provided the cut you plan to have for dinner tonight. There is one outstanding

TIMETABLE FOR THAWING FROZEN MEATS

(Time variations are influenced by degree of cold in the refrigerator or in the room where meat is thawed.)

Cut	*Thawing in Refrigerator*	*Thawing at Room Temperature*	*Thawing in Room in Front of Fan*
Beef			
Rib roast (rolled or standing)	8-10 hrs. per lb.	3-5 hrs. per lb.	1-2 hrs. per lb.
Chuck or rump roast	7-9 hrs. per lb.	2-3 hrs. per lb.	1-1½ hrs. per lb.
Steaks			
1 in. thick	8-10 hrs. ea.	3-5 hrs. ea.	1-1½ hrs. ea.
½ in. thick	4-6 hrs. ea.	2-4 hrs. ea.	¾-1 hr. ea.
Ground beef			
1 lb. pkg.	10-16 hrs.	3-5 hrs.	1-1½ hrs.
Individual patties	4-6 hrs.	1-2 hrs.	½-¾ hr.
Stewing cubes (1 lb. pkg.)	12-18 hrs.	3-5 hrs.	1-1½ hrs.
Veal			
Roast (leg, loin or shoulder)	6-10 hrs. per lb.	3-5 hrs. per lb.	1-2 hrs. per lb.
Chops, ¾ in.	4-6 hrs.	3-4 hrs.	1-1½ hrs.
Cutlets, ½ in.	3-5 hrs.	2-3 hrs.	¾-1 hr.
Stewing cubes (1 lb. pkg.)	12-18 hrs.	3-5 hrs.	1-1½ hrs.
Lamb			
Whole leg	8-10 hrs. per lb.	4-6 hrs. per lb.	1-2 hrs. per lb.
Roasts	7-9 hrs. per lb.	2-4 hrs. per lb.	1-2 hrs. per lb.
Chops			
¾ in.	4-6 hrs.	3-4 hrs.	1-1½ hrs.
½ in.	3-5 hrs.	2-3 hrs.	½-¾ hr.
Stewing cubes (1 lb. pkg.)	12-18 hrs.	3-5 hrs.	1-1½ hrs.
Pork (Because of longer cooking time required for pork and ham, cooking may be started while the meat is still partially frozen if desired.)			
Roast (loin, shoulder or center)	8-10 hrs. per lb.	3-5 hrs. per lb.	1-2 hrs. per lb.
Chops			
1 in.	4-6 hrs. ea.	3-4 hrs. ea.	1-1½ hrs. ea.
½ in.	3-5 hrs. ea.	2-3 hrs. ea.	½-¾ hr. ea.
Sausage meat (1 lb. pkg.)	10-16 hrs.	3-5 hrs.	1-2 hrs.
Sausage patties	4-6 hrs.	1-2 hrs.	½-¾ hr.
Bacon (½ lb. pkg.)	1-2 hrs.	½-¾ hr.	¼-½ hr.

COOKING METHODS AND TIMETABLE FOR THAWED OR FROZEN MEAT

Cut	Cooking Method *	Cooking Time Required: Thawed	From Frozen State
Beef			
Standing rib roast	Roast at 300°	*Rare:* 18-20 min. per lb.	38-45 min. per lb.
		Medium: 22-25 min. per lb.	42-50 min. per lb.
		Well: 27-32 min. per lb.	47-65 min. per lb.
Rolled rib roast	Roast at 300°	*Rare:* 25-30 min. per lb.	48-55 min. per lb.
		Medium: 32-35 min. per lb.	52-60 min. per lb.
		Well: 37-45 min. per lb.	55-70 min. per lb.
Round or rump roast	Roast at 300° or braise	25-30 min. per lb.	45-55 min. per lb.
Pot roast: arm, blade, chuck, brisket	Braise, cook in liquid	25-30 min. per lb.	45-55 min. per lb.
		2-inch steaks	
Steaks: Sirloin, Porterhouse, T-bone, Club (Delmonico)	Broil	*Rare:* 30-40 min.	45-50 min.
		Medium: 50-70 min.	60-80 min.
	Panbroil	*Rare:* 15-20 min.	23-25 min.
		Medium: 25-35 min.	30-40 min.
		1-inch steaks	
	Broil	*Rare:* 15-20 min.	20-25 min.
		Medium: 20-30 min.	30-40 min.
	Panbroil	*Rare:* 8-12 min.	12-15 min.
		Medium: 12-15 min.	15-20 min.
		½-inch steaks	
Top-round steaks	Broil	*Rare:* 8-10 min.	10-12 min.
		Medium: 10-12 min.	12-15 min.
	Panbroil or panfry	*Rare:* 4-5 min.	8-10 min.
		Medium: 6-8 min.	12-15 min.
Swiss steak	Braise	2 hrs.	3 hrs.
Stew	Cook in liquid	2 hrs.	3 hrs.

COOKING METHODS AND TIMETABLE FOR THAWED OR FROZEN MEAT (*Continued*)

Cut	*Cooking Method* *	*Cooking Time Required* *Thawed*	*From Frozen State*
		1 inch high	
Hamburger patties	Broil	*Rare:* 12-15 min.	20-25 min.
		Medium: 15-20 min.	25-30 min.
		½ inch high	
		Rare: 6-8 min.	10-14 min.
		Medium: 8-10 min.	12-15 min.
		1 inch high	
	Panbroil	*Rare:* 6-8 min.	10-14 min.
		Medium: 8-10 min.	12-15 min.
		½ inch high	
		Rare: 3-4 min.	6-8 min.
		Medium: 4-5 min.	8-10 min.
Meat loaf	Roast at 300°	*Rare:* 25 min. per lb.	
		Medium: 30 min. per lb.	
Veal			
Leg (shank or rump)	Roast at 300°	25 min. per lb.	40-50 min. per lb.
Loin	Roast at 300°	30-35 min. per lb.	50-60 min. per lb.
Shoulder			
Rolled	Roast at 300°	40-45 min. per lb.	55-65 min. per lb.
Cushion	Roast at 300°	30-35 min. per lb.	50-60 min. per lb.
Breast			
Rolled	Roast at 300°	40-45 min. per lb.	50-60 min. per lb.
	Braised	Total, 1½-2 hrs.	Total, 2-3 hrs.
Stuffed	Roast at 300°	40-45 min. per lb.	
	Braise	Total, 1½-2 hrs.	
Chops, ¾ in.	Braise	45-60 min.	55-65 min.
Cutlets, ½ in.	Braise	35-50 min.	45-55 min.
Stew	Cook in liquid	2-2½ hrs.	2½-3 hrs.
Lamb			
Leg	Roast at 300°	30-35 min. per lb.	40-45 min. per lb.
Shoulder			
Whole	Roast at 300°	30-35 min. per lb.	40-45 min. per lb.
Rolled	Roast at 300°	40-45 min. per lb.	50-55 min. per lb.
Cushion	Roast at 300°	30-35 min. per lb.	40-45 min. per lb.

COOKING METHODS AND TIMETABLE FOR THAWED OR FROZEN MEAT (*Continued*)

Cut	*Cooking Method* *	*Cooking Time Required* *Thawed*	*From Frozen State*
Breast			
Rolled	Roast at 300°	30-35 min. per lb.	40-45 min. per lb.
	Braise	Total, 1½-2 hrs.	Total, 2-2½ hrs.
Stuffed	Roast at 300°	30-35 min. per lb.	
	Braise	Total, 1½-2 hrs.	
Chops			
2 in.	Broil	22 min.	26-30 min.
1½ in.	Broil	18 min.	20-22 min.
1 in.	Broil	12 min.	15-18 min.
Stew	Cook in liquid	1½-2 hrs.	2-3 hrs.
Patties			
1 in.	Broil	15-18 min.	20-25 min.
	Panbroil or panfry	8-10 min.	10-12 min.
½ in.	Broil	8-10 min.	12-15 min.
	Panbroil or panfry	4-5 min.	6-8 min.
Pork			
Loin			
Center	Roast at 300°-350°	35-40 min. per lb.	45-50 min. per lb.
Whole	Roast at 300°-350°	15-20 min. per lb.	25-35 min. per lb.
End	Roast at 300°-350°	45-50 min. per lb.	60-70 min. per lb.
Shoulder			
Rolled	Roast at 300°-350°	40-45 min. per lb.	50-60 min. per lb.
Cushion	Roast at 300°-350°	35-40 min. per lb.	45-55 min. per lb.
Chops	Braise	45-60 min.	55-65 min.
Spareribs	Roast at 350°	30-35 min. per lb.	40-45 min. per lb.
	Braise	Total, 1½ hrs.	Total, 2 hrs.
	Cook in liquid	30 min. per lb.	40 min. per lb.
Fresh ham	Roast at 300°-350°	30-35 min. per lb.	45-55 min. per lb.
Butt	Roast at 300°-350°	40-45 min. per lb.	50-60 min. per lb.
Smoked Hams			
Whole			
Large	Roast at 300°-350°	15-20 min. per lb.	25-35 min. per lb.

COOKING METHODS AND TIMETABLE FOR THAWED OR FROZEN MEAT (Continued)

Cut	*Cooking Method* *	*Cooking Time Required*	
		Thawed	*From Frozen State*
Medium	Roast at 300°-350°	18-25 min. per lb.	35-45 min. per lb.
Small, or half	Roast at 300°-350°	22-26 min. per lb.	35-45 min. per lb.
Picnic	Roast at 300°-350°	35 min. per lb.	45 min. per lb.
	Cook in liquid	35-45 min. per lb.	45-55 min. per lb.
Shoulder butt	Roast at 300°-350°	35 min. per lb.	45 min. per lb.
Ham steaks			
1 in.	Broil	20-30 min.	30-45 min.
½ in.	Broil	15-20 min.	25-35 min.
Variety Meats			

NOTE: Because by far the best results are obtained from thoroughly slow-thawed or nearly thawed variety meats, no times will be given here for cooking them from the frozen state.

When method calls for braising, this manner of cooking may be done on top of stove or in a 300° oven.

Liver		
Beef	Braise	20-25 min.
Calf	Broil	8-10 min.
Pork	Braise	20-25 min.
Lamb	Broil	8-10 min.
Kidney		
Beef	Cook in liquid	1-1½ hrs.
Veal, Lamb, Pork	Broil	10-12 min.
	Cook in liquid	45-60 min.
Heart		
Beef		
Whole	Braise, or cook in liquid	3-4 hrs.
Sliced, or stew	Braise, or cook in liquid	1½-2 hrs.
Veal, whole; Pork, whole; Lamb, whole	Braise, or cook in liquid	2½-3 hrs.
Tongue		
Beef	Cook in liquid	3-4 hrs.
Veal	Cook in liquid	2-3 hrs.

COOKING METHODS AND TIMETABLE FOR THAWED OR FROZEN MEAT (*Continued*)

Sweetbreads or Brains	1. Pre-cook in liquid for 15-20 min., then broil from 10-15 min. additional 2. Braise for 20-25 min.
Tripe, beef	Simmer in liquid for 1-1½ hrs., then broil for 10-15 min. if desired

* Methods of cooking are those recommended by the home economics department of the National Live Stock and Meat Board.

exception to this generalization: Variety meats, especially liver, are richer than muscle meats in B vitamins and are exceptionally rich in Vitamin A.

From the standpoint of palatability and tenderness, however, meat varies from portions you can cut with a fork to those which are as tough as synthetic rubber.

The goal of all meat cookery, of course, is to bring whatever cut you are preparing to the optimum level of tender succulence and flavor. Whether or not you arrive at that goal will depend on how aptly you suit the cooking method to the cut of meat, for I am assuming that when you made your purchase you knew, having read the section on recognizing meat quality, the age, grade and texture expectancy of the cuts you froze, and labeled the packages accurately.

Fine-grained meat cut from the animal's involuntary muscles (such as the loin) is likely to be tender, especially if it comes from young creatures with good conformation. Well-exercised muscles, such as those of the neck, legs, shoulders and around the joints, develop rather thickened cell walls and tissues and therefore require more cooking to render them tender unless you are prepared to do a lot of chewing.

It has been said, even by me, that freezing tenderizes meat. Authorities refute this statement, saying that there is no scientific data to support it. I still think it does, however; at least a little; at least when frozen meat is thawed completely and cooked immediately. This may of course be just another example of organoleptic judgment, based on nothing more realistic than my own fancy. But whether or not freezing tenderizes meat, there is seldom any reason or excuse for serving tough, rubbery main courses. Proper cooking methods and accurate timing can make even the sitting end of a bad-tempered bull a pleasure to eat.

Roasting

This method is reserved for tender cuts of beef, veal, lamb or pork. Properly, this is cooking with *dry, slow heat,* and latest scientific experiments have eliminated former practices of searing and basting. High temperatures cause excessive shrinking, uneven cooking and loss of juiciness.

Searing is not recommended because, contrary to past belief, seared meat loses more juices than meat which is permitted to brown slowly.

Basting has been discarded as unnecessary, as roasts placed in a pan or rack with the fat side uppermost baste themselves as they cook.

Roasting should never be done with any kind of a cover. If meat is covered its escaping moisture is trapped within the pan and the piece of meat, no longer a roast, is cooked by moist heat.

The addition of water to the pan in which roasts are cooking has a tendency to decrease the meat's juiciness. The major reason why many housewives put water in the bottom of a roast pan is to prevent extreme darkening of the drippings intended for gravy. When meat is cooked at the recommended low temperatures, however, the drippings do not scorch to dark brown.

Many cooks, especially those who have learned the culinary art within the last decade, rely on a meat thermometer to determine accurate doneness rather than on the minutes-per-pound guide. There are several types of meat thermometer, all of which have been designed to register the internal temperature of a roast at its centermost part, and each cut or variety of meat has its predetermined temperature reading for degrees of doneness. For example:

Beef roasts in a 300° oven are cooked rare at thermometer reading 140°, medium at thermometer reading 160°, and well done at thermometer reading 170°.

Veal roasts, which should always be well done, are ready when the meat thermometer reading is 170°.

Lamb is roasted to perfection at thermometer reading 175°-180°.

Pork, which may be roasted at either 300° or 350°, should be allowed to remain in the oven until its interior temperature registers 185°.

Processed smoked ham, baked at 300°, is ready to eat at 160° except picnic hams and smoked butts, which should be 170°.

Despite all the scientific data at hand to tell you when a roast has had enough cooking and may be removed from the oven, there is still another phenomenon to consider if what you are after is the perfect roast—done, as the saying goes, to a turn. (I rather imagine that this expression harks back to the original means of roasting—on a turning spit in front of a fire.) Unless the roast is cut into immediately, the very instant after it is removed from the oven, it will continue to cook in its own heat! The National Live Stock and Meat Board informs me that it is very possible for a standing roast of beef to be taken from the oven when a meat thermometer registers 140°, or rare—and, sitting on a platter waiting to be carved, cook itself medium-done in a matter of minutes. Remember this, the next time your family dawdles on its way to the Sunday dinner table, and don't let them blame you if the beef is less pink than they like it.

Broiling

This method is reserved for tender (or tenderized) steaks of beef, usually not thicker than two inches, lamb chops and patties, ham slices and bacon.

Like roasting, broiling means cooking uncovered in dry heat—either over hot coals, under a gas flame or in an electric broiling unit, including the infra-red type.

Also like roasting, certain previously accepted ideas have been discarded in favor of newer broiling experiments. Many cooks still believe that broiling should be done at a high temperature in order to endow the surfaces of a steak or chop with a dark crustiness while keeping the interior more or less rare. Newer information has demonstrated, however, that moderately low broiling temperatures are preferred in order to cook the steak uniformly and insure its greater tenderness. In addition, moderate broiling temperatures prevent excessive shrinkage, eliminate smoking and charring, and consume less fuel. If you can control your broiler heat, the recommended temperature is 350°F. If your broiler is permanently set, put rack farther away from the heat source.

Veal and fresh pork are never broiled by accomplished cooks; veal because its limited fat content and abundant connective tissue require long, slow cooking in moist heat for tenderness, and pork because thorough cooking is needed to bring out the maximum of flavor.

Panbroiling

This differs from broiling in that the heat is indirect, being transmitted from the hot metal or flame-proof glass of a frying pan instead of from the fire itself. Panbroiled meat need not be rubbed with fat unless it is extremely lean. The fat within the meat will cook out and should be poured off as it accumulates. The meat should not be covered.

Braising

This method, recommended for less tender cuts of beef and for pork and veal chops, means essentially browning the meat in a small amount of added fat, then covering it and cooking it slowly in its own juices or in a little added liquid. For example, pot roasts are cooked by braising; so is a ham steak which is browned in a pan on top of the stove and then placed in a casserole to simmer in milk or fruit juice.

For braising, many of the less tender beef slices (flank, bottom round, etc.) are occasionally pounded with a spiked wooden mallet, the side of a plate or the back of a heavy knife. This is called *scoring,* and its purpose is to shorten the fibers and connective tissue in order to achieve a state of tenderness without overcooking the meat.

Frying

This is not the best of all possible ways to cook meat, even thin steaks or chops, but it is widely practiced nevertheless. The method resembles braising until after the meat has been browned, but it differs thereafter because frying is done in an *uncovered pan.* When meat is dredged with flour or bread-crumbs and fried at low heat, the result differs from braising also in that the surface becomes crisply browned.

Cooking in Liquid means making a stew.

Poultry from the Freezer

Unless you are preparing a chicken stew, fricassee or soup, do please take it out of the freezer in plenty of time for it to thaw completely in the refrigerator. The results are so very much better that it is well worth planning ahead. If time catches up with you, thaw the bird at room temperature. If even this amount of time is too much, why not have something else for dinner? Roasting or frying chicken is so far superior when cooked after complete thawing that I am not even going to give cooking times for cooking it from the frozen state. In a pinch, small broilers packaged so that they are easily separated into halves, may be broiled without thawing.

TIMETABLE FOR THAWING FROZEN POULTRY

Type	*Thawing in Refrigerator*	*Thawing at Room Temperature*	*Thawing in Room in Front of Fan*
Roasting chicken, duck, game bird or turkey	6 hrs. per lb.	2-3 hrs. per lb.	30-45 min. per lb.
Broiler halves	3-4 hrs. per lb.	1½-2 hrs. per lb.	20-45 min. per lb.
Cut-up parts	2 hrs. per lb.	1-1½ hrs. per lb.	15-30 min. per lb.

TIMETABLE FOR COOKING THAWED POULTRY

	Weight	*Oven Temperature*	*Time*
Roast chicken	3½-4 lbs.	350°	2-2¾ hrs.
	4½-5 lbs.	325°	2½-3 hrs.
	5½-6 lbs.	325°	3-3½ hrs.
Broiled chicken	Small	350°	40 min.
	Medium	350°	50 min.
	Large	350°	60 min. (Turn broiler off and continue to cook for 10 min.)
Fried chicken	Pieces	Brown in hot fat, then cover and put in 300° oven.	30 min.
Roast duck			
Domestic		325°	20-30 min. per lb.
Wild		325°	*Rare:* 10-12 min. per lb. *Medium:* 15-20 min. per lb.
Hunter's choice		500°	20 minutes total
Turkey	8-13 lbs.	325°	3½-4 hrs.
	14-18 lbs.	300°	4-5 hrs.

Fish and Shellfish from the Freezer

To be its best, fish is thawed while still in its wrappings in the refrigerator, because this method decreases the loss of flavor-bearing liquids. Allow from 8 to 10 hours per pound of packaged fish, from 12 to 18 hours for a pint package of seafood frozen in liquid.

Eggs from the Freezer

These are most satisfactory if thawed on the refrigerator shelf, but they are also serviceable if they have been packed in watertight containers which you can place in cool water until the eggs become liquid enough to use.

Thaw *dairy foods—butter, cream and cheese*—in the refrigerator until they have arrived at the proper consistency for use.

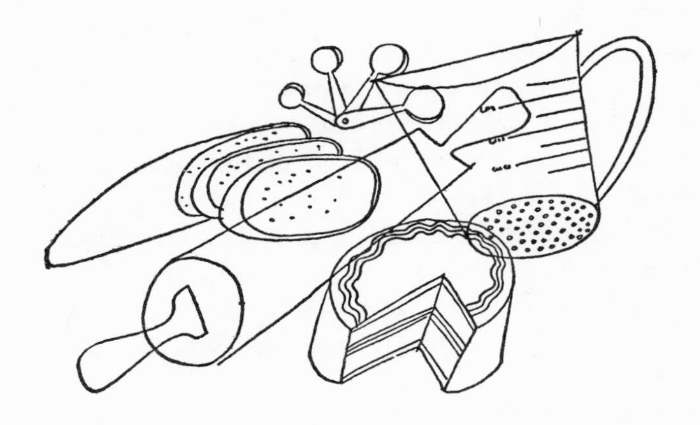

21. Baker's Dozens

Before I owned a freezer, I used to regard women who bake their families' bread and rolls with approximately the same awe and admiration I hold for geniuses who paint, sculpture, compose music or do arithmetic in their heads. Once in a while, bravely, I would buy a box of prepared roll mix, follow the printed directions with scowling intensity and turn out a dozen or so cloverleaf rolls. That, however, was the extent of my "baking."

Still, I was haunted by tantalizing fantasies of home-baked bread, for there is nothing to compare with crusty, golden loaves brought warm and fragrant from the oven. The flannel texture of most commercially mass-produced bread annoys me, and even bakeshop products leave something to be desired.

I decided to bake a loaf of bread, and did. It was delicious, and it lasted for exactly two meals. Emboldened by my initial success, but resenting the inordinate amount of time bread-baking consumes for such brief enjoyment, I set aside a day and made dozens of loaves of white bread, whole wheat bread, rye bread and cornmeal bread.

From that day, I have not bought a loaf of anybody's pale, pasty, paper-wrapped bread no matter how "enriched" it is supposed to be. I do my own enriching as I go along, using whole grains or unbleached flour (when I can find it) and adding extra milk solids and occasionally some wheat germ to my recipes.

Dizzy with confidence, I branched out. I made banana bread, orange bread, nut bread and sweet fruit bread.

It takes very little more time, and no more fuel, to make six or twelve loaves instead of one. And the great joy of it all is—my home-baked breads (rolls, too) will stay as fresh as new-baked in my freezer for as long as a year!

You have a wide and handsome choice in the matter of freezing breads and rolls as well as cakes, cookies, muffins, biscuits, pies, tarts and all other baked or batter-cooked foods. They may be frozen fully baked, half baked or unbaked. You may freeze dough, sponges or batter. There are certain advantages and disadvantages in all of these methods, and your final decision will be made on the basis of convenience plus a certain aesthetic consideration.

Freezing Dough and Batter

Once, when ambition and appetite loomed larger than my good sense, I started a bread-baking session at 10 o'clock at night. By the time I had measured, sifted, mixed, kneaded and set the bowls of dough to rise, I began doubting my wisdom. At the end of the first rising, I no longer doubted. I needed bed more than bread. Three large bowlfuls of ballooning dough looked reproaches at me, but fatigue made me ruthless. I punched them down, formed six duck-pin size balls and wrapped each one separately in aluminum foil. I put them on the bottom shelf of my freezer, intending to remove them for thawing, second rising and baking the following day.

My intentions were good, but on the following day I was sent on assignment to another city, and remained away for two weeks. When I returned, I finished the baking job. The bread was excellent, I must say, but it took forever for that dough to thaw and rise, even though I started it in a slow oven before shaping the loaves.

As a recommendation, I would say that yeast bread doughs should be frozen unbaked only in emergency or if for some reason you want the beautiful smell of baking bread in your house at a later date, instead of today. Punch them down after the first rising and press them into *thin bricks* or disks (no higher than two inches) before packaging separately in foil, cellophane or freezer paper. My duck-pin balls were a mistake. Smaller, flatter masses of dough will thaw faster.

Muffin, waffle and pancake batters can be frozen, if you like, but usually it takes just as long to complete them after freezing as it does when you start from scratch. It's a good thing to know that they *can* be frozen, however, in case you ever overestimate the capacity of a waffle or pancake crowd.

Freeze muffin batter in paper baking shells which are inserted in muffin tins, ready for baking. Wrap cellophane, foil or freezer paper around the tins. Let them thaw either completely or partially before baking. Package waffle or pancake batter in airtight rigid containers. This, of course, must be wholly thawed before you can use it. Actually, it's a much better idea to go ahead and make the waffles or pancakes and freeze them when they are cool by slip-sheeting with double thicknesses of cellophane. When you are ready to use them, just heat them in a pop-up toaster or in the oven.

Freezing Half-baked Rolls

It's a little too risky to do this with bread, but yeast rolls can be shaped and baked in a moderate oven (300°) for about 15 minutes, until they are raised but not brown. Let them cool completely, then either stack them in a large rigid container or, if your freezer can accommodate a bulky package without squashing it, put them in a double polyethylene bag. When you want to serve them, take as many out as you need and finish the baking process in a hot oven (400°) for another 10 minutes, or until they are as brown as you like them.

Freezing Baked Bread and Rolls

This is the method I almost invariably use, barring emergencies. It's simple, and there's always a supply of fresh bread and rolls on hand without further ado. Just allow them to cool, wrap bread well in cellophane, foil or freezer paper and put rolls in double polyethylene bags, and freeze. Rolls can later be heated in a roll warmer or in a slow oven. Enough bread for the meal can usually be sliced—with a good, serrated bread knife—before the loaf thaws, and the uncut portion returned to the freezer for another day.

Of course, you need not limit your freezer storage space to home-made products. Baker's and wrapped breads stay fresh, too.

As a matter of fact, day-old bread, offered by some chain stores and bakeries at a substantial discount, will emerge from the freezer almost as tasty and fresh as if fresh-baked. And, when you're traveling in a part of the country which specializes in regional baking (Cape Cod's wonderful Portuguese bread, for example) why not buy up a dozen or so loaves for future enjoyment?

RECOMMENDED MAXIMUM STORAGE PERIODS FOR BREAD, ROLLS, ETC.

Bread dough, yeast, unbaked	2 weeks
Bread, fully baked	12 months
Rolls	
Unbaked	2 weeks
Half baked	12 months
Fully baked	12-15 months
Muffins	
Unbaked	2 weeks
Baked	3 months
Waffle and pancake batter	2-4 weeks
Waffles and pancakes	6 months

Freezing Cakes and Cookies

Here, too, the choice is yours to make. Generally speaking, there is no real advantage in freezing cake batters, for they will stay fresh in the freezer as fully baked cakes up to 12 months, depending on the type (see chart to follow). Cakes frozen as batter are likely to shrink and you may run the risk of having them sink or fall later on.

An excellent idea to stockpile ready desserts in the freezer is to bake dozens of layers of all kinds—angel, sponge, white, chocolate, ginger, fruit—and keep them wrapped and sealed in a covered ice-cream tub for decorating as the need arises. Fillings and frostings are best added just before serving anyhow, although leftover filled and decorated cakes are successfully frozen if care is taken in packaging them to prevent mashing in the freezer. If you feel that you may be pressed for time when a cake party is in the offing, you might prepare filling and frosting a few days beforehand, and freeze them in airtight rigid containers.

Almost any of your favorite fillings and frostings can be frozen in this way, although the cream types may not be as satisfactory as the candy, buttery or fruity ones. Don't attempt to keep cream-filled leftover cakes too long in freezer storage, as they spoil rather readily. All bought cakes also freeze well, within the limitations mentioned.

Cooky Dough, unlike cake batter, keeps very well in the freezer for long periods of time, and may even be preferable in some instances. Some short and crisp types of cooky, if frozen baked, may be so fragile that they break no matter how carefully you handle and package them.

Package cooky dough by forming it into a cylindrical roll of the desired diameter and wrapping it in cellophane, laminated freezer paper or aluminum foil. To protect it against flattening or dents in the freezer, you can outer-wrap the roll in corrugated paperboard and then in kraft paper. For holiday and gift cookies, I save tubular containers such as ginger-snap cartons and the cardboard rolls that calendars come in. If you want square cookies, you can pack the dough down in rinsed-out milk cartons. Be sure to seal everything.

This type of freezer-cooky dough can usually be sliced while still frozen. If it is too hard for even your best and sharpest knife, let it soften just a little by putting it—wrapped—on a refrigerator shelf for about 45 minutes to an hour. Don't let it get too soft, or it will be hard to slice.

Doughs for drop or fancy-cut cookies are very conveniently packed in rigid plastic or waxed tub containers, or even in shoulderless glass freezer jars. These must thaw, of course, before they can be manipulated.

Baked Cookies, after they are thoroughly cool, may be packed in any good airtight box or carton between layers of cellophane. The largest sizes of the lined Thermorex containers are fine for cookies, especially if there are children in the house, because they require no outer wrapping. To safeguard freshness and minimize breakage, fit crumpled-up pieces of waxed paper or cellophane in any air spaces.

Freezing Pies

Miss Faith Fenton, home economist at Cornell University, has pointed out that freezing pies is not a new idea. "Our grandmothers baked several weeks' winter supply of mince pies and froze them on the back porch or in the attic, then thawed them out, or reheated them, or baked them as they were needed," she wrote in Cornell's extension bulletin #692, "Foods from the Freezer."

Opinions differ as to whether pies should be frozen baked or unbaked, and a sensible idea might be to try it both ways before deciding which way you are going. On the one hand, *unbaked* pies—especially those with fruit fillings—seem to have a fresher aroma and flavor and the pastry shells are crisper and flakier; on the other hand, *baked* pies can be stored for much longer periods of time.

Fruit, berry and mince pies are the best freezables, pumpkin and squash next, with cream or custard types last on the list but not least, for these also freeze well.

Baked or unbaked, a pie may be frozen in the metal, foil or oven-proof paper pie plate in which it was prepared. Put it in contact with a freezer surface before wrapping and storing. After it is solidly frozen, it can be slipped into a waxed pie box prudently saved from a store-bought frozen pie and relabeled as to its new content. For long storage, it is wise to wrap the carton in freezer paper, also labeled, or slip it into a plastic bag. Lacking any store-bought frozen pie cartons, the pie can be covered with a second paper or foil pie plate, then wrapped and labeled.

After they are frozen and wrapped, baked or unbaked pies can be stacked right-side up in the freezer for more compact storage.

When freezing unbaked pies, do not cut or prick the top

crust until after freezer storage, when the oven has softened the crust.

RECOMMENDED MAXIMUM STORAGE PERIODS FOR CAKES, COOKIES AND PIES

Cake	
Unbaked	6 months
Baked	12 months
Angel or sponge cake, baked	6 months
Cookies	
Unbaked	6-8 months
Baked	12 months
Pies	
Fruit	
Unbaked	4 months
Baked	6-8 months
Pumpkin or squash, unbaked (preferred)	4 months
Cream or custard	
Unbaked	1 month
Baked	4 months

NOTE: It is best not to top any pie with meringue before freezing, as the beaten egg white has a tendency to toughen when frozen, and packaging is a problem, too. Use your (defrosted) frozen egg whites—or fresh ones—to make meringue just before serving.

22. Recipes for the Freezer

ENTERTAINING MADE EASY

Whether you are confronted with an intimate late supper for six, an impressive (you hope) dinner for the boss and his wife or party refreshments for an unruly mob, your freezer can perform for you the duties of a hired caterer, relieving you of party panic and last-minute misgivings. Even unexpected drop-in guests need never throw you into a tizzy if your freezer is stocked with a few prepared main courses, plenty of interesting sauces for improvised meals and extra containers of your own beloved culinary specialties.

Of course, one of the easiest things in the world, if you're rich and reckless, is to rush to the freezer and snatch out a dozen Delmonico steaks or as many squabs when you see your husband shepherding a crowd of golf-playing cronies up the front pathway to your house. To be successful, however, entertaining need not be lavish or costly. I have kept a houseful of hungry party guests happy with such simple fare as spaghetti and meat balls (made with ground chuck or flank and veal and pork trimmings), a tossed green salad and black, bitter *espresso* coffee.

When you prepare pre-cooked party meals or snacks for freezing, choose recipes which normally are time-consuming or tedious, rather than those you can whip up quickly. It makes good sense, for example, to freeze a large quantity of meat balls, a fancy casserole or several dozen assorted open-face tea sandwiches, whereas the spaghetti should be cooked immediately before serving (leftover spaghetti freezes well, however).

Keep in mind that freezing is likely to intensify the flavor of herbs and seasonings, and go a little lighter than usual when preparing dishes expressly for frigid storage. More seasonings can be added in most cases to foods as they are heated for serving.

The few recipes suggested in the pages to follow by no means imply any limits to the kinds and quantities of cooked or ready-prepared foods you can freeze; they are intended only to indicate your freezer's versatility and to offer a guide to the variety of entertainment fare available to you. Practically all of your favorites can be successfully frozen, excepting only the crisp salad vegetables served raw and recipes using the whites of hard-boiled eggs, which toughen under extremely low temperatures.

Both unfrozen and frozen foods can be used in all recipes. The rule about refreezing thawed foods does not apply to those which are taken from the freezer raw, cooked and cooled. *The cooking process changes the enzymic quality of food.* When you remove and thaw a beef roast, for example, then cook it according to your customary method, it can be considered as an entirely new food and frozen for storage up to six months.

"Blue-Heaven" Hamburgers (for 24)

8 pounds of ground beef (let thaw, if frozen)
1 loaf of white bread, without crusts
4 teaspoons salt
1 teaspoon black pepper

---

1 pound bleu cheese (or use Roquefort, to be fancy)
4 tablespoons Worcestershire sauce
1 teaspoon dry mustard
½ teaspoon tabasco sauce

Crumble the bread and moisten it slightly with milk or water, then work it into the beef with salt and pepper. Blend the cheese with seasonings. Divide the meat into twelve parts and make four patties from each part, 48 in all. In the center of 24 patties put about 1½ tablespoonfuls of the bleu cheese mixture, then top with the remaining patties and pinch the edges together firmly. Broil under moderate heat for about 3 minutes for each side. Allow patties to cool.

Package the hamburgers in containers by slip-sheeting with double thicknesses of cellophane to prevent them from sticking together. Remove from the freezer at least six hours before you plan to serve them so that they may thaw partially in the refrigerator. Broil the patties for another 3 minutes on each side to complete cooking, and serve on toasted buns.

(If you make this recipe with unfrozen beef, freeze the patties after they are filled, without cooking, and broil them when they are thawed, 6 minutes on each side.)

Meatballs in Sauce (for 24—serve with spaghetti)

4 pounds of ground beef
1 pound of ground lean pork
1 pound of ground veal
6 large cloves of garlic
3 cups of minced (finely chopped) onion

2 cups of dry bread crumbs or 3 cups of loosely packed fresh bread crumbs
1 cup olive or salad oil
1 cup milk
3 whole eggs
1 tablespoon salt
½ teaspoon black pepper

---

3 quarts frozen or canned whole tomatoes
2 cups water
½ cup lemon juice
1 teaspoon salt
1 teaspoon dry basil
1 teaspoon dry oregano
1 teaspoon black pepper
½ teaspoon cayenne pepper

(Unless you have a simply enormous skillet, it would be better to divide the meat mixture into batches after you have prepared it, and brown one part at a time.)

Mash garlic or put it through a garlic press directly into a large skillet, then add onions and oil. Mix together the ground beef, pork and veal with bread crumbs, milk, lightly beaten eggs, salt and pepper. Form into balls (about 75). Brown meatballs in the skillet.

In a large pot, put the tomatoes, water and lemon juice and add the dry seasonings and herbs. Mix well, and start cooking over low heat. Lift meatballs and onions from oil with a slotted spoon and add them to the tomato mixture. If you and your guests like garlicky food, add a whole peeled clove or two to the liquid at this time, but don't forget to fish for it later.

Simmer for two hours, then remove from flame and let it cool. Package by putting as many meatballs as you can, without crushing them, in large tub-type or plastic containers, and cover to within ½ inch of the top with sauce.

When you are ready to serve, it is not necessary to defrost

this recipe. Just turn it out in a pan and heat. If the sauce seems to be too thick, add only enough water (or tomato juice) to reduce it to a proper consistency.

Baked Beans (for 24)

Real home-baked beans are a rare treat, these days, and your assembled guests will be delighted with them, especially if they are served with old-fashioned brown bread. To be the genuine article, they must be started from scratch, tended lovingly and baked for a long, long time. Use your favorite traditional recipe, or follow the one below. If you have brown earthenware bean pots, the large size, don't freeze the beans in them unless you plan to serve a whole potful at one time. It is better to transfer the baked beans into quart-size waxed tub or plastic containers for freezing.

3 pounds of dried beans (marrowfat, navy or pinto)
1½ pounds of lean salt pork
3 medium onions
3 cups dark molasses or 3 cups brown sugar
3 teaspoons salt
3 teaspoons dry mustard

Wash beans in colander, then pick them over and discard any doubtful ones. *Soak beans overnight* in water to cover. Cook them the next day in the soaking water to which you have added enough more to cover them. Let them come to a boil, then lower the heat and simmer them gently until blowing on a bean cracks its skin. Remove from fire.

For this quantity of beans, you will need three two-quart bean pots (or heavy aluminum pots) with covers. Place one whole onion at the bottom of each pot. Divide cooked beans, molasses and seasonings into three parts and, after they are well mixed, put one-third of the mixture into each of the bean pots. Cut the salt pork into 2-inch cubes and tuck the pieces into the beans near the top. Add water to fill the pots to within an inch of the tops, cover, and bake in a slow oven (300°) for 10 hours. Add water to fill pots again at the end of 3 hours, and again at the end of 6 hours from the start. After the beans have baked for 10 hours, remove the pots' covers and bake for another hour.

Big and Little Pizzas

In recent years, a new word has flashed across our highways in neon lights. Where once the weary road traveler saw, mile after mile, the word "Eats," "Diner," "Hamburgers" and "Red Hots," he now also sees—and almost as frequently—"*Pizzeria*" or, more simply, "*Pizza.*"

This good Italian specialty is becoming as familiar to us as spaghetti, ravioli and other culinary contributions of the sunny peninsula. To make it properly, with Neapolitan or Sicilian authenticity, you would need an enormous brick charcoal oven and a long-handled wooden shovel. Americanized variations, however, make very good late-party conversation pieces, especially when served with beer or dry red wine or, more circumspectly, Italian demitasse.

The following recipe will make a pie (English word for "*pizza*") 11 inches in diameter, or enough for 4 to 6 servings.

Use your own favorite yeast bread dough recipe or 2 cups of packaged hot roll mix, prepared according to directions on the box.

Let dough rise until double in bulk (about 1 hour) then pound or roll into a very thin round, turning up the edge about ½ inch.

Brush surface of unbaked shell with 2 tablespoons of olive oil.

Drain a pint of (thawed) frozen or canned whole tomatoes, cut them into pieces and spread evenly over the dough.

Sprinkle with scant ½ teaspoon of dry oregano (or basil) and with a little salt and black pepper. If you like, you may also sprinkle with discreet amounts of cayenne powder or crushed red pepper.

Cut into small pieces about ½ pound of Mozzarelle cheese (if this is not available to you, use Muenster or Cheddar, although these are not nearly so good) and spread over the entire surface of the pie.

Bake in hot oven (400°) for about 5 minutes, not enough to melt the cheese thoroughly but sufficient to stiffen the dough.

Let cool, package by covering with a paper pie plate and wrapping in cellophane or freezer paper, and freeze.

When you are ready to serve your *pizza,* dot the pie with your choice of cut-up anchovy fillets, capers, button mushrooms, small pieces of Italian sausage, or leave it plain. Bake in a hot oven for 10 minutes, then place the pie under the broiler to melt and brown the cheese.

Cut into wedges and serve with a small bowl of grated Parmesan cheese.

Little pizzas: Use rounds of ordinary white bread or thin slices of long loaves of French or Italian bread.

Toast bread lightly on one side, under the broiler.

Brush untoasted side with olive oil and spread with canned Italian tomato paste, then dot with small pieces of Mozzarelle cheese.

For added variety and piquancy, dot several groups of the little *pizzas* with one of the following: sliced stuffed olives, capers, minced or pressed garlic, chopped anchovy fillets or curls, minced Italian sausage, crumbled liverwurst, sliced cooked mushrooms, chopped cooked shrimp.

Package the assorted *pizzas* by putting them on cooky

sheets and covering with foil or laminated freezer paper. Freeze.

When you are ready to serve them, sprinkle with olive oil, a little oregano or crushed red peppers (vary the seasonings) and put under a low broil flame until the cheese is melted and lightly brown.

One loaf of bread will make approximately 20 *pizzas,* and each 20 will use up a 6-ounce can of tomato paste.

Hot Rollwiches

Use thin-sliced sandwich bread.

Trim crusts, then sprinkle each slice of bread lightly with water. Flatten slices between sheets of waxed paper with a rolling pin, being careful not to tear the bread.

Spread the slices with any of the following: minced ham, liverwurst, Cheddar cheese, chopped tongue and pickle relish, drained flaked tuna fish, flaked crab or lobster meat, mashed boneless and skinless sardines, peanut butter and crisp bacon crumbs, or any sandwich spread which does not contain egg white or salad vegetables.

Roll the spread bread into fingers and secure with toothpicks. Broil under a low flame until light brown. Remove toothpicks and package in containers or polyethylene bags after the rollwiches have cooled. Freeze. To serve, return them to the oven until they are heated through.

(The same idea can be carried out a little more impressively by using thin French-style pancakes instead of plain bread.)

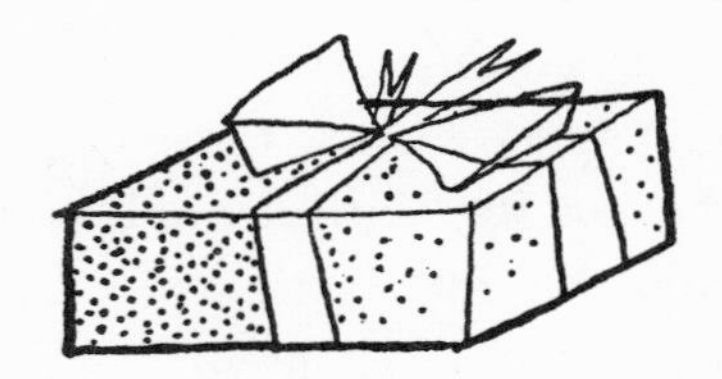

FROZEN GIFTS SHOW A WARM HEART

The gift of food is a welcome gift, from the slab of fresh-baked cake sent to welcome a new neighbor to the bowl of nourishing chicken broth brought to cheer an ailing friend.

Your freezer offers many opportunities for gracious giving. Most people, I have found, are warmed and touched by the gift of a gaily wrapped fruit cake or jar of special cookies at Christmas time.

Throughout the year, birthday and anniversary remembrances of an especially liked recipe are received with gladness and appreciation of your thoughtfulness.

More than one of my friends, convalescent at home after a boring regime of prophylactic hospital food, has been almost tearfully grateful for the gift of a choice two-inch steak (raw) or a whole broiled small chicken instead of a bottle of cologne or another bed jacket.

Cookies and candies, frozen separately and later packaged in beach pails or colorful toy carts, are always highly regarded by children, whose parents are equally pleased with special breads and cakes.

The selected recipes which follow are only a few of those you can make at your leisure, store in the freezer and later package appropriately for holiday or special-occasion gifts.

Popcorn Balls (about 20 medium-sized balls)

Pop 2½ cups of dry popping corn to get 12 cups of popped corn.

Combine 1 cup of dark molasses, 1 cup of corn syrup and 1 tablespoon white vinegar. Cook these ingredients over low heat, stirring occasionally, until a small amount of the mixture dropped from the end of a spoon forms a very hard, brittle ball in cold water. Remove from heat and add 3 tablespoons of butter or margarine. Pour molasses mixture over the 12 cups of popped corn and mix well. Allow it to cool enough to touch.

With wet or lightly buttered hands, roll and mold the popcorn into balls. Cool them on waxed paper. When they are cool, wrap each ball in foil, then in Christmas tissue paper or colored cellophane, twisting the ends. Package for the freezer in large polyethylene bags and store them in the cabinet so that they will not be crushed.

Christmas Balls (about 50 small balls)

Chop together 1 cup of puffed breakfast cereal (rice, wheat or corn) with 1 cup each of seedless raisins, dates and figs. Mix in 1 cup of chopped nut meats. Add your choice of ½ cup of either candied orange peel or mince meat. Mix these ingredients with 2 teaspoons of vanilla or maple extract. Form into balls and roll in one or several of the following: shredded coconut, cocoa, shaved milk chocolate, cinnamon, powdered ginger, crushed peppermint candy or decorettes. Wrap in decorative paper or colored cellophane, leaving long twisted ends. Package for freezing in large polyethylene bags. For gifts, gather long ends of paper together to form bouquets, necklaces or bunches of "grapes."

Freezer Kisses

Make a quantity of basic fondant. The following recipe will yield approximately one pound. To make more, simply multiply each of the ingredients by the number of pounds you wish to have.

2 cups of sugar (white or brown)
1¼ cups water
2 tablespoons light corn syrup

Cook these ingredients together in a saucepan over low heat, stirring until the sugar is completely dissolved. Continue to cook without stirring, removing any crystals which form on sides of pan with a wet cloth tied around the bowl of a long wooden spoon. Cook until the mixture has reached the soft-ball stage, when a small amount dropped off the end of a spoon will form a soft ball in cold water. Remove at once from the heat, and pour out on a large china platter which you have dipped in water. Let cool to lukewarm, then beat with a spatula or wooden spoon until the fondant becomes creamy and white. At this stage, work the fondant with your hands, kneading until it becomes smooth. Put the fondant in a large jar or bowl, cover it, and let it stand in your refrigerator for at least 24 hours before proceeding with the following steps.

To make kisses, melt fondant in top of double boiler over hot water. If you are making a variety of flavors, melt only enough at one time for a particular flavoring, repeating the procedure for each variation.

When the fondant is melted, add one of the following to each batch: shredded coconut, broken nut meats, candied fruits, crushed peppermint sticks or oil of peppermint, crushed cinnamon drops, maple extract, shaved chocolate, etc.

Cut cellophane or heavy aluminum foil into 4-inch squares and drop the candy from the tip of a teaspoon in the center of the paper. The candy should hold its shape. If it appears too soft, allow the mixture to cool before dropping it.

Twist the candy-filled paper into kisses and store in freezer boxes or in polyethylene bags.

Christmas Cookies

Prepare your favorite gingerbread or sweet cooky dough, roll it out to the desired thickness and cut or shape it into people, animals, Christmas trees and other fancy shapes. Freeze them unbaked between double thicknesses of cellophane or aluminum foil on cooky trays. When frozen, stack them carefully in trunk-opening freezer boxes, separated by cellophane strips, until you are ready to bake them. Bake unthawed in a moderate oven until they are done, then decorate them with icing, bits of candied fruit, raisins, nut meats, etc.

Chocolate Brownies (about 40 two-inch squares)

4 squares of baking chocolate
1 cup of shortening (solid)
2 cups of sugar, white or brown
4 eggs
1 cup all-purpose flour
½ teaspoon salt
2 teaspoons vanilla
1 teaspoon baking powder
2 cups broken nut meats

Melt shortening and chocolate together in top of double boiler, mix well, and turn into a bowl. Add sugar, and mix until dissolved. Add the well-beaten eggs and vanilla. Sift together the flour, salt and baking powder and stir into mixture. Fold in nut meats. Turn batter into two 10-inch square pans, well greased, and bake in moderate oven (350°) for 35 minutes. Cut into two-inch squares before removing from pan. When cool, package each brownie separately in cellophane, or package in freezer boxes collectively by separating each layer with double thicknesses of cellophane. Individual brownies may be gathered together and frozen in polyethylene bags until you are ready to gift-wrap them.

Date and Nut Bread

This recipe makes two small loaves. To make more—and you'll surely want to—multiply all ingredients by number desired.

1 cup of chopped pitted dates
1 cup of chopped nut meats
½ cup sugar, white or brown
1 teaspoon baking soda
2 beaten eggs
2 tablespoons melted butter or margarine
1¾ cups all-purpose flour
1 teaspoon vanilla

Soak chopped dates in 1 cup of boiled water, allow to cool in large bowl. To the soaking dates, add soda, beaten eggs, butter or margarine, flour and vanilla. Fold in nut meats. Mix thoroughly. Turn into two small loaf pans, well greased, and bake in moderate oven (350°) for 1 hour. When cool, wrap each loaf separately in aluminum foil for freezing.

Banana Bread

Here's a chance to use those frozen mashed bananas!

This recipe makes two 9-inch loaves of delicious banana-flavored bread.

3½ cups of sifted all-purpose flour
4 teaspoons baking powder
½ teaspoon baking soda

1 teaspoon salt
⅔ cup solid shortening
1⅓ cups granulated sugar (or substitute brown sugar)
4 well-beaten whole eggs
1 pint (2 cups) mashed bananas (if frozen, let thaw until workable)

Sift together flour, baking powder, soda and salt. Paddle shortening with a wooden spoon until it is creamy and fluffy. Add sugar to softened shortening gradually, continuing to work the mixture until it is light. Add beaten eggs and whip them well into the mixture. Slowly add sifted flour mixture and bananas, alternating the two and beating smooth after each addition. Turn into greased 9-inch loaf pans and bake at 350° for 70 minutes.

When cool, wrap loaves in cellophane or aluminum foil and freeze.

Frozen Eggnog

Freeze a batch for your own holiday callers, of course, but what nicer gift can you take to Christmas hosts and hostesses than a decorative jugful of the traditional wassail, perhaps accompanied by a few whole nutmegs wrapped like small Christmas balls?

The following recipe will make one gallon:

10 egg yolks
3½ cups white sugar
1½ cups Bourbon (or rum or brandy, as you prefer)
10 egg whites
1 quart whipping cream

Beat egg yolks and sugar together until light and creamy. Add liquor to mixture and stir. Beat egg whites until stiff and fold into mixture. Last, fold stiffly whipped cream into the eggnog. Spoon into wide-mouthed jugs, being sure to leave at least 1 inch expansion space at top. Cover mouths with heavy aluminum foil and seal with green and red freezer tape or tie closely with Christmas ribbon.

Remember to take the jugs out of the freezer at least six hours before you plan to serve or present them as gifts. They should not be refrozen.

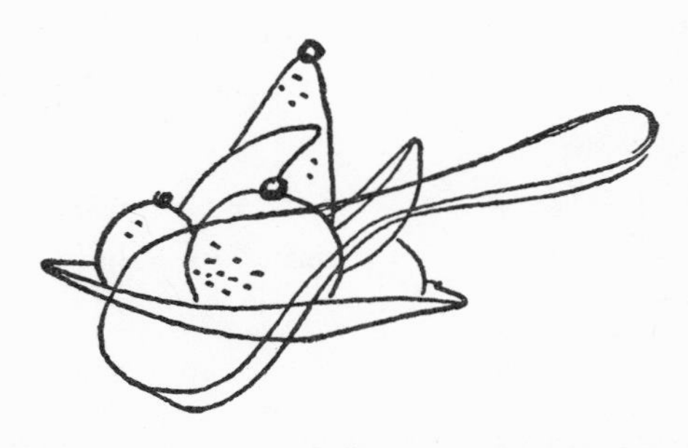

ICE CREAM . . . AND OTHER FROZEN ASSETS

Let your freezer be a fairytale chest, offering up the magic of special desserts and fancy salads.

The best ice cream is still made in the old-fashioned type of crank freezer, although the new electric ones do a fine job, too. The recipes given here, however, may be made with great success right in your refrigerator or freezer for packaging in smaller quantities.

The frozen salads are best made in individual molds or containers.

Many of the ingredients called for come from your freezer. Thaw them before using.

Avocado Sherbet

1½ cups mashed avocado
1 cup lemon juice
1 cup orange juice

2 cups sugar
1 pint heavy cream, whipped
2 teaspoons grated lemon rind

Combine the fruit juices, avocado, sugar and lemon rind, blending until sugar is completely dissolved. Pour this mixture into ice-cube trays and freeze for about ½ hour. Remove from the freezer, turn it out into a bowl and stir it well. Fold in the whipped cream. Pack in rigid waxed tub or plastic containers and store in freezer. Twelve servings.

Pumpkin Ice Cream

1½ cups frozen (or canned) pumpkin pulp
1½ cups sugar
2 cups milk
4 egg yolks
2 tablespoons cornstarch
1 teaspoon cinnamon
½ teaspoon each of ginger, mace and salt
1 teaspoon grated orange rind
2 teaspoons unflavored gelatine
¼ cup cold water
1 cup broken nut meats
1 cup whipped cream

Dissolve gelatine in ¼ cup of cold water and set aside. Combine sugar, cornstarch, spices and salt. Scald milk and add all dry ingredients, cooking over low heat until thickened. Add beaten egg yolks, pumpkin, orange rind and dissolved gelatine, stirring until well blended. Pour into ice-cube trays and chill, do not freeze. When well chilled, remove from trays and fold in whipped cream and nut meats. Package in rigid containers and freeze. Twelve servings.

Peanut Brittle Ice Cream

2 teaspoons unflavored gelatine
½ cup cold water
1¾ cups evaporated milk (scald)
½ cup sugar
2 teaspoons vanilla or 1 teaspoon maple extract
1½ cups heavy cream, whipped
¼ pound peanut brittle, crushed into small pieces

Soften gelatine in cold water, then dissolve it in the scalded evaporated milk. Add sugar and extract, stirring until sugar is dissolved, then allow this mixture to cool. Pour into ice-cube trays and chill in the refrigerator until slightly thickened. Fold in whipped cream and return to ice-cube trays, then freeze for 1 hour, until the mixture has the consistency of slush. Pour into a chilled bowl and beat vigorously until it is smooth, but not melted. Fold peanut brittle into the mixture. Package in rigid containers and freeze. One quart.

Chocolate Ice Cream

2 teaspoons unflavored gelatine
½ cup cold water
1¾ cups evaporated milk (scald)
¾ cups sugar
2 squares unsweetened baking chocolate
1 teaspoon vanilla
1½ cups heavy cream, whipped

Soften gelatine in cold water; then dissolve chocolate in the scalded evaporated milk, beating it with a rotary beater or in an electric mixer until thoroughly blended. Add all other ingredients except whipped cream, and allow to cool. Pour into ice-cube trays and chill in refrigerator until slightly thickened. Fold in whipped cream and return to ice-cube trays. Freeze for 1 hour, until the mixture has the consistency of slush. Pour into a chilled bowl and beat vigorously until it is smooth, but not melted. Package in rigid containers and freeze. One quart.

Coffee Ice Cream

Substitute ½ cup of extremely strong coffee for the chocolate squares, and proceed exactly as for Chocolate Ice Cream.

Berry or Fruit Ice Cream

2 teaspoons gelatine
½ cup cold water
1¾ cups evaporated milk (scald)
½ cup sugar
2 teaspoons vanilla
1½ cups heavy cream, whipped
1 pint container any frozen berry or fruit (sweetened) except bananas

Drain berries or fruit, after thawing, until very little liquid remains. Soften gelatine in cold water and dissolve it in the scalded evaporated milk. Add sugar and vanilla, and allow it to cool. Pour into ice-cube trays and chill until slightly thickened. Fold in whipped cream and return to ice-cube trays, then freeze for 1 hour, until the mixture has the consistency of slush. Pour into a chilled bowl and beat vigorously until smooth, but not melted. Fold berries or fruit into mixture. Package in rigid containers and freeze. One to one and a half quarts.

Banana Ice Cream

Allow 1 cup of frozen banana pulp to thaw until it has reached the stage when you can stir in ¼ cup of lemon juice. Follow procedure for Berry Ice Cream, adding the banana and lemon mixture before the final freezing.

Frozen Strawberry Salad

2 cups (1 pint container) frozen sliced strawberries, partially thawed and drained
12 ounces cream-style cottage cheese (or smooth cream cheese)
½ cup heavy cream, whipped

Combine strawberries with cheese until well blended. Fold in the whipped cream. Package in *waxed tub containers* and freeze. To serve, peel the carton from the cylindrical frozen salad and slice. Bed on lettuce leaves and top with mayonnaise.

This recipe will make two pint tub containers, or one quart container. Other sliced or crushed berries may be substituted for strawberries.

Frozen Fruit and Nut Salad

4 cups (2 pint containers) of any diced fruit, fresh or frozen (thawed)
⅓ cup mayonnaise
½ teaspoon salt
1 cup chopped nut meats (may use peanuts)
1½ cups diced marshmallows
1½ cups cream, whipped

Mix all ingredients, folding in whipped cream. Pour mixture into 12 individual waxed paper cups and top each one with a berry or a cube of fruit. Freeze. When frozen, wrap each cup in cellophane or aluminum foil and store in polyethylene bags. Serve frozen, removed from cups, on a bed of crisp lettuce leaves. Top with a whipped mixture of mayonnaise and heavy cream, sprinkled with nutmeg.

Cranberry Ribbon Salad

1 pint container frozen whole cranberry sauce (or 1 pound can)

2 tablespoons lemon juice
1 cup heavy cream, whipped
¼ cup powdered sugar
1 teaspoon vanilla
¾ cup chopped nut meats

Stir lemon juice into whole cranberry sauce and pour into ice-cube tray. Combine whipped cream, sugar, vanilla and nut meats and spoon this mixture over the cranberry layer. Freeze in tray, then wrap tray in cellophane or aluminum foil.

To serve, remove from tray and slice, bedding each piece on crisp lettuce. Eight servings.

Frozen Cinnamon Apples

6 apples, tart variety
1 cup sugar
1 cup water
2 cups hard candy, cinnamon flavor (red)
3 ounces cream cheese
⅓ cup mayonnaise
½ cup chopped nut meats

Add sugar and cinnamon candy to water and simmer until both are dissolved. Pare and core apples and simmer them in the candy syrup until they are almost tender. Cool. Blend together the cream cheese, mayonnaise and nut meats and fill centers of apples. Freeze in individual waxed tub containers, or quick-freeze on cooky sheet and package later in cellophane or aluminum foil.

23. Random Facts, Fancies and Foibles

TAKING DREARINESS AND DRUDGERY OUT OF SPECIAL DIETS

When someone in the family (or a frequent visitor) is on a special diet, your freezer can help considerably in removing a large part of the extra effort it always takes to prepare recipes and menus for the dieter. Your freezer can, moreover, make dieting less of a bore through the simple expedient of stockpiling properly controlled foods and meals for interesting variety.

If, for example, the diet is a reducing one—high in proteins and low in carbohydrates and fats—you can make life pleasanter for the dieter by reserving a special section in the freezer for a collection of portion-controlled lean steaks, chops, poultry, fish and variety meats which can be popped under a broiler, either frozen or thawed, and spiced with allowable herbs and seasonings.

Go even further in your sympathetic understanding of a dieter's conflict with monotony, and prepare in advance a week of varied dinners, for dinner is the meal which most usually demoralizes the would-be stoic. Even the most wistful or rebellious meal-martyr is likely to feel less abused if each night's dinner offers a different appetizer, main course, vegetable and dessert.

Grapefruit or orange sections, mixed fruits, melon balls, frosty whole peeled tomatoes and clear soups provide variety at the beginning of a meal, sugarless desserts can brighten up its end, and all can come out of the freezer with a minimum of effort for the housewife who is also preparing menus for the non-dieters in her life. Resolute dieters who are unhappy without sweets can be appeased if you package containers of fruits and berries covered with syrup made of saccharine instead of sugar. Just remember that 1 quarter-grain saccharine tablet is equal in sweetening power to 1 heaping teaspoonful of sugar. (This is a hint for diabetic diets, also.)

Cottage cheese, allowable on most reducing diets, freezes extremely well. Rather than serve up plain cottage cheese for lunch, however, buy it in bulk and repackage it in small meal-size containers, mixing each portion with one of a number of palatables such as chopped pimiento or green pepper, fresh pineapple, unsweetened berries, finely minced cucumber and onion, chives, carroway or poppy seeds or chopped tomatoes.

Those on salt-free or low-salt diets need no longer submit to foods which have the savor of sawdust, for all other seasonings except parsley and celery flakes have recently been found to contain so little sodium that they can be used with safety. When freezing vegetables for such dieters, therefore, add discreet quantities of basil, oregano, rosemary, tarragon or marjoram to the blanching water. During freezer storage, the flavors will be intensified to provide tastier meals. These and other herbs can also be added to the broiled or roasted main courses recommended for low-salt diets.

The salads which most dieters must consume daily can be enlivened by adding, to the unfreezable raw greens, slivers of freezable cooked (lean) pork, chicken, fish and skim-milk cheese. Keep small quantities of these on hand in the freezer, packaged in little polyethylene bags.

Because unflavored gelatine is largely protein, it can be called into service to provide a variety of diet luncheons. Freeze gelatine molds containing small quantities of mixed vegetables, lean bits of meat and cooked shellfish or diced fruit to be served later on lettuce leaves with a low-calory salad dressing. (Remember to drop pineapple into boiling water for a minute or so if you plan to use it in gelatine.)

Baked potatoes for dieters may be frozen a dozen or so at a time if you remove the potato from the skins and mash it with skim milk, returning the pulp to the skins and freezing each half packaged in cellophane, foil or laminated freezer paper. Heat without thawing in a slow oven and sprinkle with paprika before serving.

Second to sweets, bread is the nemesis of even strong-minded people on a diet. Most regimens allow one slice of bread or its equivalent per meal, however, and the dieter can be made to feel cherished and less alone in the world if you will take the little effort it requires to bake several loaves of nutritious whole-wheat bread or dozens of whole-grain muffins, especially if you use some of last summer's blueberries, currants or unsweetened fruit in the muffin batter. Baked breads will keep beautifully in freezer storage for as long as a year, you know, so a day's baking can reap dividends of appreciation for many months.

All in all, dieting for whatever purpose is a nuisance and a frustrating experience, for those who must diet are usually those to whom eating has been a joy. Deprived of gastronomic delights, the dieter in your home can nevertheless enjoy the compensations of little attentions and prettily prepared plain foods. A freezer makes it all so simple.

LUNCHBOX HUNCHES

Have you any idea how many sandwiches you make during a year? A mother with two school-children who carry their lunches makes around 800 sandwiches a year. If her husband also carries a lunchbox, she makes up to 600 more. Then, too, there are picnic sandwiches, after-the-movies snacks and quick pick-up lunches for herself as she works around the house.

Your freezer can be of tremendous help to you in preparing school or work lunches, for, with proper planning, you can make a two- or three-week supply of sandwiches and other box fillers in advance, freeing you from last-minute flurries and daily decisions about what in the world to prepare for Johnnie's or Susie's lunch tomorrow.

More important, however, is the fact that the freezer will permit you to plan in advance and so be able to provide better balanced nutrition for your children. Doctors tell us that there is no such thing as a "balanced meal." Each day must provide a day's nutritive value in the foods we eat. A school child, therefore, must receive approximately one-third of the entire day's food value from the lunch carried to school or bought in the school cafeteria.

A satisfying lunch for a child should be one which feeds his body while filling his stomach; and if his lunchbox contains nourishing juices and sweets he will be less inclined to squander his hard-earned allowance money on sugary soda pops and candies which displace vital food elements and attack his teeth.

From the freezer, for example, come fried chicken legs and breasts and hearty meat pasties (see recipe on page 379) which offer a welcome change from the sandwich course from time to time. Packaged in colorful muffin cups or in little half-pint plastic containers and removed from the freezer in the morning, salads will be thawed to the just-right temperature when the lunch bell rings. And, while not all soups taste good when cold, several do.

Frozen fruit juices, whole wheat cookies and fruity candies made with honey for sweetening are nutritional positives, and so are the containers of chocolate milk which can be mixed beforehand in quantity and frozen in individual containers for lunchbox toting.

In your favorite cookbooks you will find many recipes for soups which are delicious when served cold. These are the ones to make for school lunchboxes, freezing them in little 8-ounce glass, plastic or waxed tub containers. A few suggestions are: Vichysoisse, tomato bouillon, cream of avocado, spinach, asparagus, beet and cranberry borscht.

When making sandwiches, be sure to take advantage of the variety of breads available. Whole wheat, rye, pumpernickel, graham and fruit-flavored breads keep as well in freezer storage as white bread, and so do any number of roll varieties. You will find, too, that bread used right from the freezer is easier to spread than unfrozen slices.

When preparing sandwich spreads or fillings for freezer storage, avoid lettuce and the whites of hard-boiled eggs. Get away from the rut of cheese-and-ham or liverwurst by trying some of the following:

Chopped frankfurter (cooked) with pickle relish, mustard and chili sauce
Chopped chicken with nuts and Russian dressing
Ground ham with chopped dill pickles and mustard
Tuna fish with chopped celery and green peppers
Mashed boneless sardine with hard-boiled egg yolks and lemon juice
Chopped Cheddar cheese and green peppers
Deviled corned beef or ham with chopped stuffed olives
Crisp bacon chopped with celery and mixed with peanut butter and honey
Meat loaf with chopped pimientos
Minced shrimp, lobster or crab with cream cheese or lemon butter
Chopped chicken or calves' livers with hard-boiled egg yolk and minced onion
Thinly sliced roast lamb with mint jelly
Cream cheese, sliced stuffed olives and crushed peanuts
Mashed avocado with anchovy paste
Chopped tongue with grated yellow cheese and mustard pickles

Wrap all sandwiches individually, cutting them first into halves, quarters or fancy shapes if you like. Use cellophane instead of waxed paper for individual wrappings, then gather up a dozen or so of identical sandwiches and package them in airtight boxes or polyethylene bags, being sure to identify the contents. Sandwiches packed into a lunchbox in the morning will be thawed and ready to eat, tasting fresh-made, by lunchtime. A few leaves of lettuce can be packaged separately in a small film bag or piece of waxed paper for the luncher to add to the sandwiches when they are thawed.

Frozen fruits, berries, freezable salads and aspics, individually packed in small moisture-vapor-proof containers as single servings, can also be tucked into school lunchboxes as pleasant—and healthful—surprises.

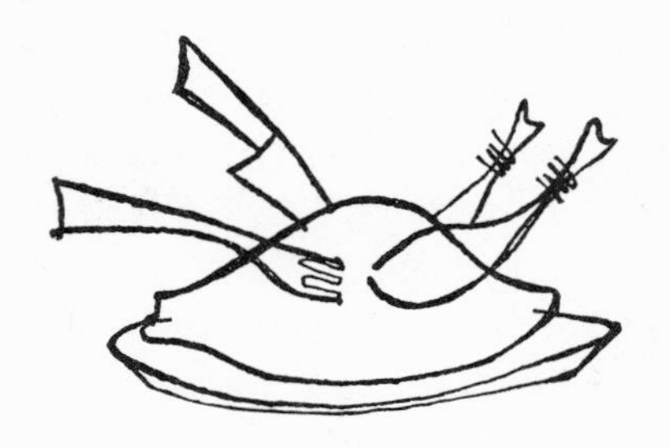

HOLIDAYS WITHOUT HYSTERICS

Remember those hectic holiday mornings when you whirled around in the kitchen while your children clamored underfoot and your husband dropped ashes all over the freshly mopped floor, and you tensed yourself for the ring of the doorbell announcing the arrival of a dozen hungry relatives?

Not any more, thank you! You, with your freezer to help you, can sail through the happy day without one degree of raised temperature. To be sure, despite the seductive ads, you won't be lolling around in a chaise longue while dinner cooks and serves itself; but, with planning, your freezer can stand as a buffer between you and holiday hysteria.

Take Thanksgiving, for example, or Christmas. The menu is almost traditional, and so—until now—is the kitchen drudgery necessary to achieve the proper festivity (seasonal word for food).

Plan your holiday dinner menus down to the last detail a month or six weeks before the event and, during the weeks between, prepare the various dishes when it is convenient for you to do so. Wrap, freeze and store them in a special place in your cabinet. When the holidays arrive, the greatest part of your preparations will have been done; and, if you freeze your stuffed turkey or chickens uncooked and your pies unbaked or half-baked, the unforgettable smell of roasting fowl or baking pastry will delude all comers into thinking that you've been up since dawn slaving away.

You can freeze practically anything and everything you like to serve for the holidays. Prepare your favorite recipes for the table, as if you were going to serve them tonight. Let them cool, then refrigerate them for a few hours. Wrap single-serving dishes—like fruit molds, for instance—individually. Tureen vegetables like turnips, peas, green beans or broccoli can be frozen in large containers. If your family expects creamed white onions, however, it is better to freeze the onions separately, covering them with their own juice when both it and the onions are cold. On the morning of the dinner, thaw them over low heat and reserve the liquid for gravies or soup stock. Make the cream sauce fresh, as it has a tendency to separate in the freezer.

You can freeze both stuffed-baked white and sweet potatoes. For white potatoes—bake them until done, then cut in halves and scoop out pulp, carefully, reserving the shells. Add cream, butter, salt and pepper to the potatoes and whip them thoroughly with a potato masher or an electric mixer until they are without lumps. Refill the shells and sprinkle paprika over the tops of the potatoes. Wrap each potato half in cellophane, laminated freezer paper or aluminum foil. Chill in the refrigerator, then freeze.

Sweet potatoes—omit cream, but add butter, seasonings and a little orange juice or grated orange rind to the pulp. Freeze in the same way as for white potatoes.

If frozen in foil, the potatoes may be heated in their wrappings by placing them on your oven rack and heating for about 20 minutes at 350°. If frozen in other freezer materials, unwrap them, place on the oven rack, dot with butter and bake for the same amount of time.

Candied sweet potatoes, prepared as if for the table and cooled, may be frozen right in their baking dish by covering

it with laminated freezer paper, cellophane and foil. For serving, heat in a 350° oven for about 20 minutes.

Turkey with stuffing can be frozen together, ready for roasting; and giblet gravy, made when you prepare the bird, can be frozen in shoulderless jars or rigid plastic containers. Your own favorite stuffing will do, but remember that herbs and seasonings tend to become stronger in the freezer. Use less of such spices as sage, thyme and savory, and go a little light on shortenings. Truss the bird, tying legs and wings close to the body, and wrap carefully in laminated freezer paper or in foil or cellophane outer-wrapped with stockinette. Don't forget to take the turkey out of the freezer in plenty of time for it to thaw in the refrigerator! (See table on page 343 for thawing time according to size of bird.)

Your cranberry sauce or relish, too, can be prepared far in advance and frozen either in individual molds or in a large one.

See Chapter 21 for directions for freezing pies.

ICE-CUBE TRICKS AND BREAD-PAN BRICKS

There is a need, among freezer owners, for very small moisture-vapor-proof containers to hold, for example, a quarter cup or two tablespoonfuls of something used in those quantities in frequently prepared recipes. It is wasteful to expend larger containers on such small amounts, and a nuisance to home-manufacture little envelopes of cellophane, freezer paper or aluminum foil—which, however, I have done.

Several of the polyethylene bag companies put out, for commercial use, tiny tubular lengths into which are stuffed cylindrical cheeses, bolognas and liverwursts as well as pickles, olives, nut meats and nickel candy. These are usually imprinted with the name of the food producer, and are sometimes emblazoned with colorful designs. Whenever I buy these products I try to remove the coverings carefully, in order not to rip or puncture them. I rinse them out in warm soapy water and then in many dousings of clear water, dry them thoroughly and put them into a larger bag for safekeeping. They are fine for small quantities of chopped vegetables or the flavoring ingredients for such recipes as chicken à la king.

Lacking a supply of these little bags, however, I use the ice-cube trays of my refrigerator for many freezing purposes. For example:

A little bit of leftover sandwich spread, such as deviled ham or tongue.

Chopped mixed green peppers, tomatoes, onions and celery to be used as seasoning.

The remains of a can of grated pineapple, wonderful to have on hand for a quick salad or to flavor a single baked ham slice.

The small quantities of butter, margarine or shortening used in some baking recipes.

Sandwich spreads, when I don't want to take the time to make the sandwiches.

Leftover coffee and tea, to be used for iced coffee and tea. Plain ice cubes dilute these drinks too much for my liking.

Fruit juices, for added flavor and color in carbonated beverages.

Plain water with a single berry, cherry, olive or pickled onion frozen in the center of each cube—for party drinks.

Water colored with pure vegetable dyes, also for party impressions. A shrimp cocktail surrounded by crushed pale green ice is a joy to behold.

Italian tomato paste, bought economically in restaurant-sized cans and decanted into ice-cube trays for small recipes of spaghetti sauce.

Concentrated beef, veal or chicken stock to add to gravies.

Mixed nut meats and raisins, to offer to visiting children in lieu of sugary confections. They love it.

Leftover sauces—curry, barbecue, cheese, etc.—to glamorize a solitary luncheon snatched between household or writing chores.

Trinkets, folded pictures or personal notes wrapped in cellophane and frozen in the center of pale fruit juices or plain water to surprise and amuse a shut-in child, who immediately becomes more willing to drain the glasses handed to him at frequent intervals during convalescence.

And, of course, single eggs, egg yolks or egg whites (see page 207).

Freeze whatever it is you are freezing in the ice-cube trays, then pop the cubes out and store them in polyethylene bags. It is not necessary to wrap all types of cubes individually, although it is better to do so with eggs, vegetables, butter and sauces. Juices and plain ice cubes, colored or not, can be stored as is. The extreme low temperature of the freezer keeps them from sticking together.

The ordinary bread pan, too, is useful for freezing blocks of food of all kinds. The advantages are that you conserve your supplies of rigid containers, and save space within the freezer. Some of the foods I have frozen as bread-pan "bricks" and wrapped in laminated freezer paper for storage are:

Soup. One brick serves from 4 to 6 people.

Applesauce.

Small fish fillets covered with ice water.

Bread pudding or moist cakes.

Chicken à la king or chili con carne which is to be used within a comparatively short period of time.

Rhubarb, blanched for 1 minute only to remove the stiffness so that it can be packed into the bread pan.

Water colored with pure vegetable dyes to be used in punch bowls. (Do this in round saucepans, too.)

DON'T BE AFRAID OF YOUR FREEZER!

As you grow into familiarity with your freezer, you find that somewhere along the way you have lost what seemed, in earlier days, to be a certain awe and rigid adherence to "The Rules." That big, white, gleaming guardian of perishable foods, when first established in your household, may have exerted a domineering influence over your habits and methods of preparation. It demanded—or so you thought—inflexible routine. It seemed to regard you with frigid scorn if you tentatively improvised your own ways and means in accordance with your own needs.

More than one freezer owner has confessed to feeling apologetic the first time she discovered an original and somewhat unorthodox use for her freezer. One young woman said to me, for example: "I wanted to frost some champagne glasses, so I dipped them in water and rubbed the rims with sugar and popped them into the freezer. I swear that the thing glared at me. Well, I glared right back and insisted on my rights as a free-thinking American. But, do you know, when I reached in for the frosted glasses, one of them had cracked! It took me a couple of weeks to get enough courage to try it again. I guess I showed that tyrant who's boss, because it's never dared crack another glass."

This endowment of malevolent intent to an inanimate object is, of course, purely subjective. It is best to start out at once with the idea that the freezer is the handmaiden and you the mistress.

Having established my own authority with my freezer early in the game, I have cajoled it into doing all sorts of chores it would probably be ashamed to admit at a reunion of its production-line buddies.

It keeps my coffee and cigarettes fresh.

It receives my dampened linens in a large plastic bag for a brief spell before they are ironed.

It chills a case of warm beer, in an emergency.

I like grist-ground wheat in some bread recipes for texture (and nourishment) but grist mills are hard to find in our modern communities. When, driving out into the country, the road leads past a feed mill, I am likely to stop and buy a large sack of freshly ground whole wheat flour. Mice and insects prefer this more nutritional product to the one which comes ultra-refined in prophylactic packages, and so keeping it around the house was a problem until I entrusted polyethylene bagfuls of it to the bottom shelf of my freezer.

My most ingenious use of the freezer almost landed me in trouble. A fellow music enthusiast was visiting one evening, and the conversation got around to old jazz records. We were arguing about a trumpet passage in an old recording, now a collector's item, of which I have a rare copy.

"Wait a minute," I said in the middle of a discussion that was getting nowhere. "I'll play it for you and you'll see. I think it's ready, now."

"What do you mean—you think it's ready?"

"It's in the freezer."

My friend looked at me as if he thought it was time to call the men in the white coats to come and get me, and I hastened to explain.

The record, while playable, was warped, and I had decided earlier that day to try to straighten it out. I put it over the radiator until it softened a little (it was one of the very old hard-rubber discs) and then weighted it down under heavy books until it flattened properly. It seemed perfectly logical to me to "set" the flatness in the freezer.

It worked, too.

Index

C

INDEX

P

Recipes

NOTES

NOTES

NOTES

NOTES

NOTES

NOTES

NOTES

NOTES